Baltic Polymer Symposium 2015

IOP Conference Series: Materials Science and Engineering
Volume 111

Sigulda, Latvia
16 - 18 September 2015

ISBN: 978-1-5108-2228-3
ISSN: 1757-8981

Printed from e-media with permission by:

Curran Associates, Inc.
57 Morehouse Lane
Red Hook, NY 12571

Printed by Curran Associates, Inc. (2016)

Institute of Physics
Dirac House, Temple Back
Bristol BS1 6BE UK

Phone: 44 1 17 929 7481
Fax: 44 1 17 920 0979

techtracking@iop.org

Additional copies of this publication are available from:

Curran Associates, Inc.
57 Morehouse Lane
Red Hook, NY 12571 USA
Phone: 845-758-0400
Fax: 845-758-2633
Email: curran@proceedings.com
Web: www.proceedings.com

Table of contents

Volume 111

2016

Baltic Polymer Symposium 2015
16–18 September 2015, Sigulda, Latvia

Accepted papers received: 29 December 2015
Published online: 26 February 2016

Preface

Papers

IOP Conf. Series: Materials Science and Engineering **111** (2016) 011001 doi:10.1088/1757-899X/111/1/011001

Preface

The issue contains selected papers of the XV international conference "Baltic Polymer Symposium (BPS)", which was held in Sigulda, Latvia, from 16–18 September 2015.

The conference is organized annually by the technical universities of the three Baltic states. Since the first event in 2001, the range of the topics covered by the conference has comprised all aspects of modern **Polymer Science**, including synthesis, processing, recycling, development and characterization of polymer composites as well as revealing the possibilities of nanotechnology. During its fifteen years BPS has gained considerable popularity and at the moment it is an important event not only for polymer scientists but also for researchers working in the related fields of **Material Sciences** and **Nanotechnology** due to the increasing multidisciplinary nature of the research area. Participation in BPS offers great possibilities of exposure to the wider field, learning, getting new experience and ideas, as well as founding new contacts. The symposium is an excellent chance for young researchers to improve their ability in presenting the achievements of their research to the international community.

The Baltic Polymer Symposium 2015 in Sigulda was attended by participants not only from the three Baltic states, but also by participants from Argentina, Germany, Finland, United Kingdom, Spain, Poland, Ukraine, Belarus and Russia. During the conference **4 plenary lectures** were given by world-recognized scientists and the audience was introduced to the most recent findings in the field of polymer materials science during **27 oral** and **105 poster presentations.**

IOP Conf. Series: Materials Science and Engineering **111** (2016) 011002 doi:10.1088/1757-899X/111/1/011002

Peer review statement

All papers published in this volume of *IOP Conference Series: Materials Science and Engineering* have been peer reviewed through processes administered by the proceedings Editors. Reviews were conducted by expert referees to the professional and scientific standards expected of a proceedings journal published by IOP Publishing.

IOP Conf. Series: Materials Science and Engineering **111** (2016) 012001 doi:10.1088/1757-899X/111/1/012001

Modelling of Honeycomb Core Sandwich Panels with Fibre Reinforced Plastic Facesheets and Analysing the Mechanical Properties

Raghul Krishni Narasimhan[1], Daiva Zeleniakiene[2]
[1] Department of Production Engineering, Kaunas University of Technology, Kaunas, Lithuania
[2] Department of Mechanical Engineering, Kaunas University of Technology, Kaunas, Lithuania

E-mail: raghul.krishni@ktu.edu

Abstract. Using experimentally obtained specific material properties, numerical finite element models were created, one was honeycomb core sandwich structure other neat FRP composite structure was designed and experimentally verified. The honeycomb core sandwich composite comprised facesheets from wound glass fibre and polyvinylester resin and a core from recycled paper hexagonal honeycomb impregnated with polyvinylester resin and neat FRP composite structure consisted only two thin layers of facesheets. The model was used to obtain the optimal thickness of facesheet in honeycomb core sandwich structure at which the effective strength and stiffness properties can be obtained. It was determined that thickness of the facesheets had a significant effect on stiffness properties when the length between the supports are high.

1. Introduction

Sandwich fibre reinforced polymer (FRP) composites have emerged as important material because of their high specific strength and high specific stiffness, light weight, high fatigue resistance compared to common metallic alloys. Composite materials are used in almost all aspects of industrial and commercial fields such as ships, aircrafts, and general vehicles [1-2]. Honeycomb structures are especially becoming more prevalent in the field of civil engineering where the need of high structural strength and low weight is necessary [3].

Sandwich panels consist of two thin facesheets covering the light weight core. For numerical analysis of honeycomb sandwich structure various modelling approaches are developed. The finite element analyse is one of the means used to find the approximation of global behaviour of the sandwich panels [4-5]. The high mechanical performance with minimum unit weight can be provided by fibre reinforced polymer honeycomb sandwich structure [6].

In order to increase the performance and use of honeycomb sandwich material in different applications, knowledge of the mechanical behaviour is required. This motivates to develop complex numerical models and experimental methods, which characterise the design, material models and optimizing the honeycomb sandwich panels in certain specific conditions.

The object of the investigation is the sandwich composite with facesheets made of wounded glass fibre and polyvinylester resin and core made of recycled paper hexagonal honeycomb impregnated in polyvinylester resin.

The aims of this study are to find the appropriate numerical material models and compare these models with experimental data; using obtained numerical models to determine the optimal thickness of facesheet in honeycomb sandwich structure at which the effective optimal strength and stiffness properties are obtained and increase the mechanical behaviour of sandwich structure.

2. Material and Modelling

The sandwich structure presented in Figure 1 was used for the investigation. In order to find the mechanical properties of the sandwich materials various tests were carried out according to the standards. For facesheets ISO 527, ISO 604, ISO14129 and honeycomb core ISO 844, ISO 1922 were used. The obtained mechanical properties are presented in Table 1 and Table 2. According to the obtained material properties of the honeycomb it is found that material is highly anisotropic. The average thickness of the ply was 0.7 mm and fibre volume was 43%.

Figure 1. Model of sandwich with honeycomb core. 1 – woven glass fibre and polyvinylester resin composite facesheet; 2 - recycled paper hexagonal honeycomb impregnated in polyvinylester resin.

Table 1. Mechanical properties of FRP facesheet.

Mechanical properties	**Value**	**Units**
Tension strength	645	MPa
Compression strength	248	MPa
Longitudinal young's modules E_1	37.5	GPa
Transverse young's modules E_2	7.32	GPa
Poisson's ratio v_{12}	0.28	-
Poisson's ratio v_{21}	0.05	-
Shear modules G_{12}	3.79	GPa
In plain shear strength, S_{12}	23.0	MPa

Table 2. Mechanical properties of paper honeycomb core.

Mechanical properties	**Value**	**Units**
Young's modules	10	MPa
Compression strength	0.48	MPa
Shear modules	235	MPa
Shear strength	0.64	MPa

The recycled paper hexagonal honeycomb impregnated with polyvinylester resin was used for sandwich core thickness of wall was 0.22 mm, height was 10 mm, and edges was 10mm. the model was modelled using shell element for facesheet and solid element for honeycomb core.

According to experimentally obtained data the models of honeycomb core sandwich structure (with a honeycomb) and neat FRP facesheet material (without the honeycomb) were created using finite element modelling in ANSYS 14.5 as shown in Figure 2. The bonding between the honeycomb core and the facesheet was modelled with a "glue" layer with the thickness of 0.05 mm. The properties of the glue for the numerical model were defined as the mechanical properties of synolite 8388-P-1 resin (Young's modulus and tensile strength were 3.7 GPa and 14 MPa, respectively).

Verification of facesheet material model was performed using linear analysis by simulating tension test. Previously, an experimental test was carried out. The laminate code was [±65/90], the thickness of the plies were 0.9 mm and 0.75 mm for ±65 and 90 plies, respectively. The total thickness was 2.4 mm. Stress versus strain was measured in this test and the linear dependence curve shows a good agreement with experimentally obtained curve as shown in Figure 3.

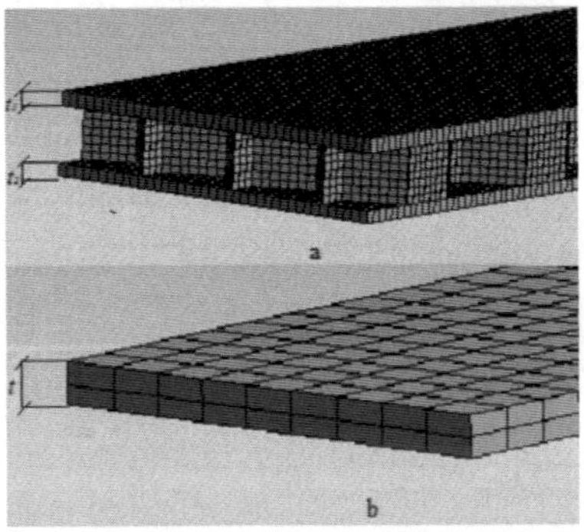

Figure 2. Models: (a) - model of honeycomb core sandwich composite; (b) – model of neat FRP sandwich composite. t_1 & t_2 - thickness of top facesheet and bottom facesheet; t ($t = t_1+t_2$) – thickness of neat FRP sandwich

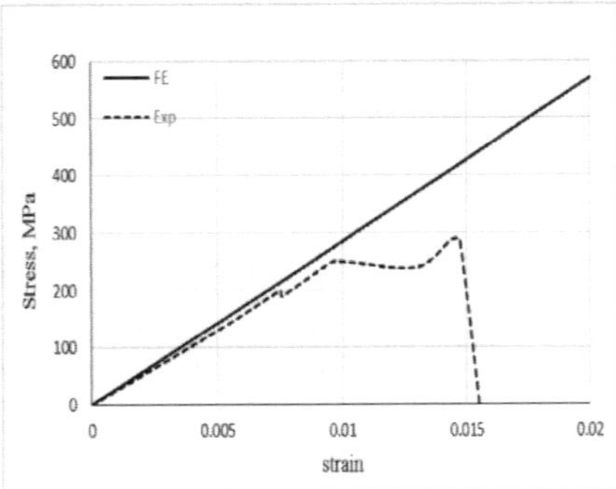

Figure 3. Tension stress – strain curve of FRP composite

Verification of the sandwich structure model was performed using linear analysis by simulating three point bending test. Previously, an experimental test was carried out. The dimensions of the specimens were as follows: width 60 mm, distance between the supports 200 mm, thickness of the top facesheet 2.68 mm, the thickness of bottom facesheet 2.81 mm, the core thickness 10 mm and thickness of sandwich 15.5 mm. The force versus deflection was measured during this test and the linear dependence curve of FE model shows a good agreement with experimental obtained curve as shown in Figure 4.

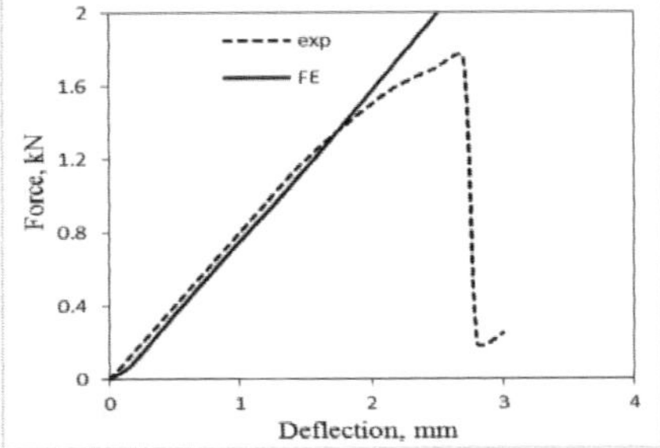

Figure 4. Force versus deflection curve obtained experimentally and by FE simulation

Using verified model the quasi-static in three point bending tests were simulated. A constant load of 800 N was applied in all simulation. This investigation was carried out in three different ways by varying the facesheet thickness. Such as $t_1 > t_2$, $t_1 < t_2$, $t_1 = t_2$ (t_1- thickness of top facesheet and t_2- thickness of bottom facesheet). In first condition $t_1 > t_2$ where, thickness of the top face sheet t_1 is varied by increasing the plys for each simulation and the bottom facesheet thickness t_2 was kept constant. In second condition $t_1 < t_2$ where thickness of the bottom facesheet t_2 was varied by increasing the plys for each simulation and top facesheet thickness t_1 was kept constant. In third condition were $t_1 = t_2$ thickness of the both facesheets t_1 and t_2 were changed equally by adding the equal number of layers for each simulation. Laminate code for the top facesheet and the bottom facesheet were changed as $[\pm 65]_n$ and $[\pm 65/90]_n$ respectively.

The thickness of facesheet was increased step by step in every investigation and three point bending simulation was simulated as shown in Figure 5. For each thickness change of facesheet, deflection and maximum equivalent stress the maximum equivalent stress was measured on middle of sandwich panel (Point where the load was applied in sandwich structure).

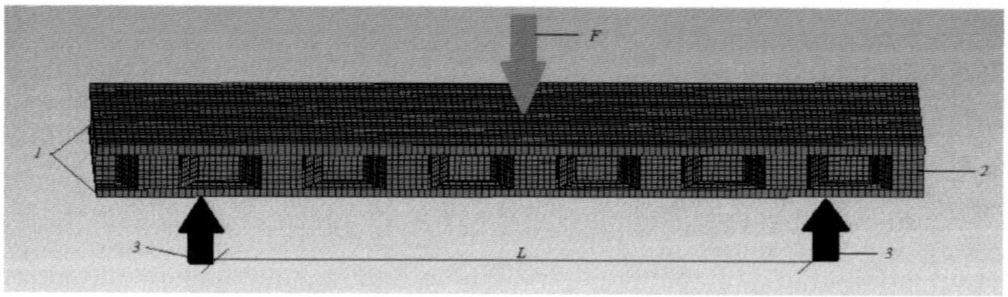

Figure 5. The numerical model of three point bending specimen. F - force applied, 1 - facesheets, 2 - hexagonal honeycomb core, 3 - supports, *L* – length between the supports.

For both honeycomb core sandwich structure and neat FRP composite the stiffness were calculated according to the equation:

$$K = \frac{F}{y_{max}} \quad (1)$$

Where K – stiffness, F – force applied, y_{max} – maximum deflection.

The maximum deflection coefficient $k_{y_{max}}$ which can be represented as ratio of maximum deflection of neat FRP composite to the maximum deflection of honeycomb core composite structure was used:

$$k_{y_{max}} = \frac{y_{max_1}}{y_{max_2}} \quad (2)$$

Where y_{max_1} - maximum deflection of neat FRP composite; $y_{max\,2}$ - maximum deflection of honeycomb core composite structure.

IOP Conf. Series: Materials Science and Engineering **111** (2016) 012001 doi:10.1088/1757-899X/111/1/012001

The coefficient $k_{\sigma_{max}}$ represented the ratio of maximum stress σ_{max} of neat FRP composite to the maximum stress of honeycomb core FRP sandwich. This can be expressed as:

$$k_{\sigma_{max}} = \frac{\sigma_{max_1}}{\sigma_{max_2}} \qquad (3)$$

Where σ_{max_1} – maximum equivalent stress of neat FRP composite; $\sigma_{max\,2}$ – maximum equivalent stress of honeycomb core composite structure.

3. Result and discussion

The stiffness variation influenced by thickness of FRP was calculated for both honeycomb core sandwich structure and neat FRP. The stiffness increases as the thickness of the FRP increases in all the cases. For lower thickness value, stiffness for honeycomb sandwich structure is higher than neat FRP composite. At a particular thickness value, stiffness of both honeycomb sandwich structure and neat FRP are same, but that particular point differs depending on the thickness orientation and distance between the supports. For $t_1 > t_2$ when $L = 100$ mm the value is $t = 9 – 10$ mm, when $L = 150$ mm the value is $t = 10 – 11$ mm, when $L = 200$ mm twice higher the other thickness. For $t_1 < t_2$ when $L = 100$ mm the value is $t = 7 – 8$ mm, when $L = 150$ mm the value is $t = 8 – 9$ mm, when $L = 200$ mm the value is twice higher the other thickness. For $t_1 = t_2$ when L=100 mm the value is $t = 8 – 9$ mm, when $L = 150$ mm the value is $t = 9 – 10$ mm, when $L = 200$ mm the value is also double. Below this thickness value, stiffness of the honeycomb core sandwich structure is higher than the neat FRP. In the same case above this thickness value, stiffness of the neat FRP is higher than the honeycomb core sandwich structure. The graph clearly represents that stiffness value of $L = 100$ mm higher than $L = 150$ mm that is more or less double the value also in $L = 200$mm.

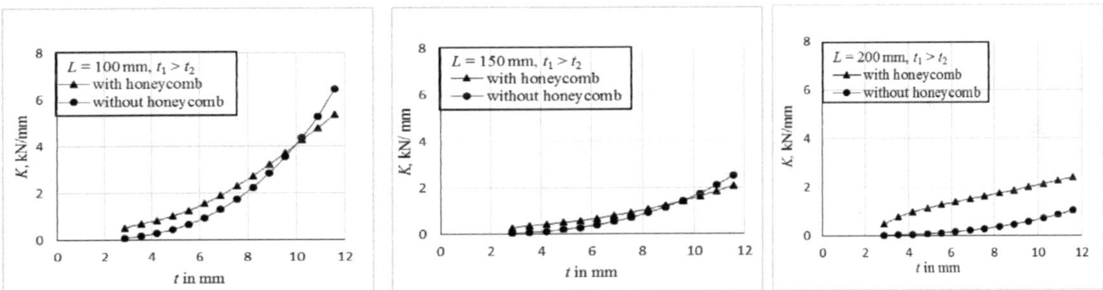

a b c

Figure 6. Influence of FRP thickness t on stiffness K: a, b, and c – where the top facesheet thickness t_1 is greater than bottom face sheet t_2 lengths between supports respectively 100, 150, 200 mm.

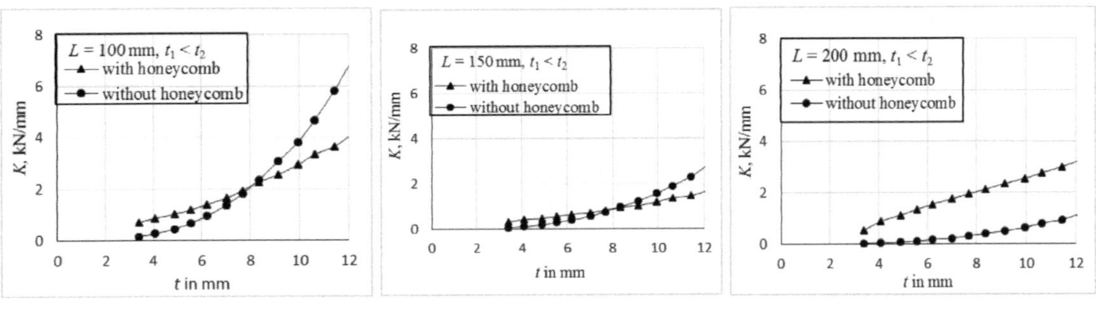

a b c

IOP Conf. Series: Materials Science and Engineering **111** (2016) 012001 doi:10.1088/1757-899X/111/1/012001

Figure 7. Influence of FRP Thickness t on stiffness K: a, b, and c – where the top facesheet thickness t_1 is lesser than bottom face sheet t_2 lengths between supports respectively 100, 150, 200 mm.

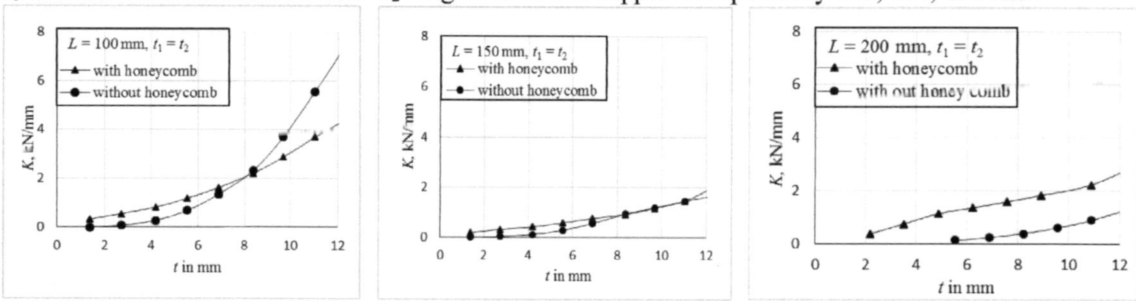

a b c

Figure 8. Influence of FRP Thickness t on stiffness K: a, b, and c – where the top facesheet thickness t_1 is equal to bottom face sheet t_2 lengths between supports respectively 100, 150, 200 mm.

By only comparing the optimal thickness, stiffness of the composite is not so clear because of the same value of stiffness can be obtained from constant F and different y_{max} values. So influence of thickness separately on y_{max} and σ_{max} were investigated.

The coefficient $k_{y_{max}}$ and $k_{\sigma_{max}}$ are defined by y_{max} and σ_{max} values, which is obtained from different thickness and length between the supports. It is clear that the deflection decreases when the thickness of the FRP increases and the distance between the support decreases. The effects of the maximum deflection value of honeycomb core composite was found only when the thickness of FRP is lower and distance between the supports is increased.

In case of neat FRP composite, it has the minutiae stiffness in the lower thickness values so coefficient $k_{y_{max}}$ cannot be calculated

In contrast for thickness t equal to 5 mm (at this value honeycomb height is 80% of the total composite thickness [7, 8]). The deflection of the honeycomb core FRP composite is close to 2.1, 2.6, and 14 times lower than the neat FRP composite in all three conditions as the distance between composite are 100, 150 and 200.

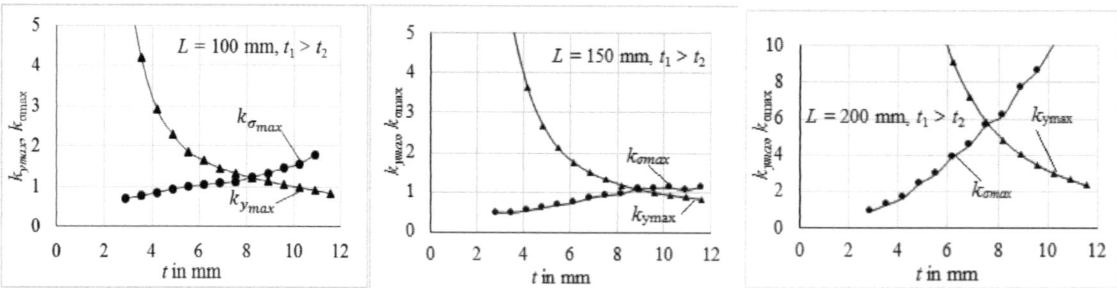

a b c

Figure 9. Influence of FRP Thickness t on coefficient $k_{y_{max}}$ and $k_{\sigma_{max}}$ - where the top facesheet thickness t_1 is greater than bottom face sheet t_2 lengths between supports respectively 100, 150, 200 mm.

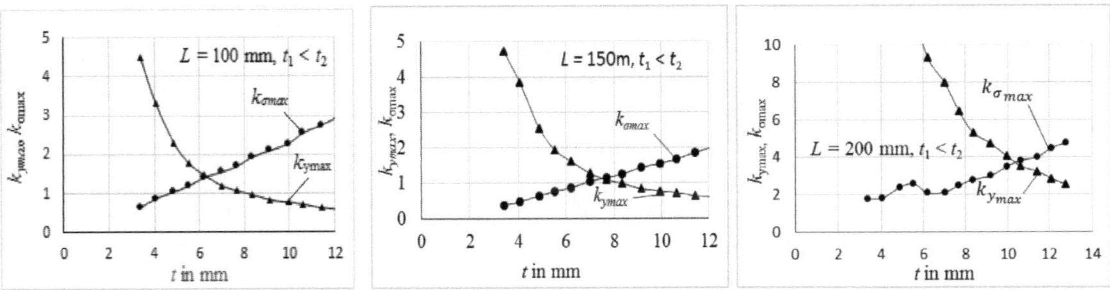

a b c

Figure 10. Influence of FRP Thickness t on coefficient $k_{y_{max}}$ and $k_{\sigma_{max}}$ – where the top facesheet thickness t_1 is lesser than bottom face sheet t_2 lengths between supports respectively 100, 150, 200 mm.

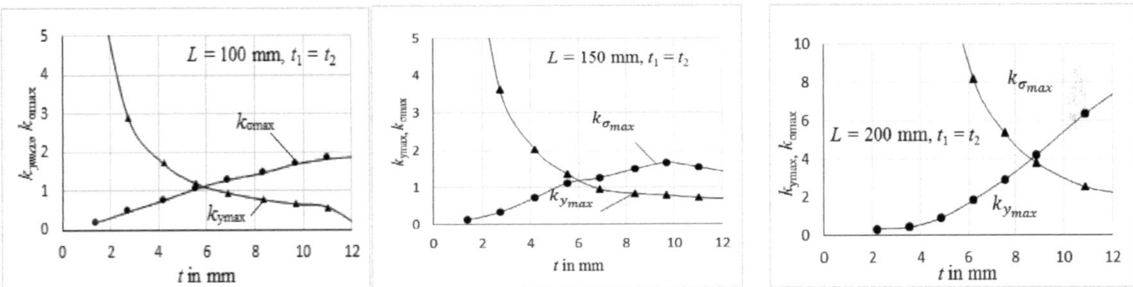

a b c

Figure 11. Influence of FRP Thickness t on coefficient $k_{y_{max}}$ and $k_{\sigma_{max}}$. where the top facesheet thickness t_1 is equal to bottom face sheet t_2 lengths between supports respectively 100, 150, 200 mm.

The significant effects of equivalent stress in honeycomb core were found only when the FRP thickness is low. It is clear that, when the thickness of FRP is increased and distance between the supports decreased, the equivalent stress in the FRP have decreased. In case of lower thickness stress on the honeycomb core FRP composite is lower than neat FRP composite in different conditions and distance between the supports. But in higher thickness value the situation is inversed and the stress on honeycomb core composite is high when compared to the neat FRP composite.

The effective performance of the honey combe core sandwich structure can be found when the coefficients $k_{y_{max}}$ and $k_{\sigma_{max}}$ are higher than one.

In the above Figure 9, Figure 10, and Figure 11 the average range of FRP thickness for all three conditions are: when L=100 mm the thickness range is 5 -9 mm, L = 150 mm thickness range is 6 – 10 mm and L = 200 mm thickness range is 5 – 14 mm, where the condition is sustain and the range of the thickness depends on the distance between the supports. When the length between the supports increased the range of thickness is also increased.

4. Conclusions

An analysis of strength and stiffness of sandwich structure comprises of honeycomb core sandwich and neat FRP were carried out.

The material of the separate components of sandwich structures were tested and the mechanical properties were obtained. Using the material properties a numerical model of sandwich structure comprising wounded glass fibre and polyvinylester resin facesheets and recycled paper hexagonal honeycomb impregnated in polyvinylester resin was modelled. The Facesheet tension test and three point bending of sandwich structure allowed to verify the FE model of facesheet material and sandwich structure.

The methodology used for investigation of the sandwich structure by changing the thickness of the facesheets in three different conditions such as $t_1 > t_2$, $t_1 < t_2$ and $t_1 = t_2$, this methodology allowed to investigating the strength and stiffness properties at various thickness of the facesheets and distance between the supports. This helped to determine the optimal thickness value of FRP in honeycomb core composite.

In result of the investigation, the optimal thickness value of FRP in honeycomb core composite was purely depends on structure geometry of material or product.

It is also equally important to consider the distance between the supports which influence the thickness variation of the FRP facesheets in honeycomb core sandwich composite.

References

[1] Wasim A, Sachin Kumar C and Syed Mazhar A 2013 *Int., J. Engg Res & Technol.* **2** - 2278

[2] Gopinatha A, Senthil Kumar M and Elayaperumalc A 2014 "*12th Global Congress on mfg., mgmt* GCMM. **97**

[3] Shen SY, Masters FJ,Upjohn HL, and Ferraro CC 2013 *Comp., Str.* **99** 419–32

[4] Meo M, Morris AJ, Vignjevic R and Marengo G 2003 *Comp. Str.* **62** 353–60

[5] Foo CC, Chai GB, and Seah LK 2007 *Comp. str.* ***80*(4)** 588-94

[6] Davalos JF, Qiao P, Xu XF, Robinson J and Barth KE 2001 *Comp. str.* ***52* (3)** 441-52

[7] Yu SD and Cleghorn WL 2005 J. snd and vib. **284** 189-204

[8] Das M, Barut A, Madenci E and Ambur DR 2005 *computer methods in appl. mechanics and engg.* **194** 2969-3005

IOP Conf. Series: Materials Science and Engineering **111** (2016) 012002 doi:10.1088/1757-899X/111/1/012002

Effect of melamine phosphate on the thermal stability and flammability of bio-based polyurethanes

Vladimir Yakushin[1], Irina Sevastyanova[1], Dzintra Vilsone[1], Andris Avots[1,2]

[1]Latvian State Institute of Wood Chemistry, 27 Dzerbenes, LV-1006, Riga, Latvia

[2]Riga Technical Univrsity, Faculty of Material Science and Applied Chemistry, Institute of Polymer Materials, 3 Paula Valdena, LV-1048, Riga, Latvia

E-mail: yakushin@edi.lv

Abstract. The effect of melamine phosphate (MP) on the thermal stability of bio-based polyurethane and the flammability parameters of wood samples with polyurethane coatings was studied. Thermogravimetric analysis and cone calorimeter test at a heat flux of 35 kW/m^2 were used for this purpose. The main characteristics of the thermal stability and flammability of the coating with addition of MP were compared with the characteristics of analogous coatings with addition of ammonium polyphosphate (APP), as well as APP in combination with melamine. It was found that the use of MP as an intumescent additive allows a considerable decrease of most of the flammability parameters of the polyurethane based on tall oil fatty acids, like APP. To reach the maximum effect, it is enough to load in the polyurethane 20% of MP. In contrast to APP, MP reduces also the smoke release of the samples. Using MP in combination with APP at definite weight ratios, it is possible to essentially reduce the flammability parameters of polyurethane coatings, such as PHRR, THR and MARHE.

1. Introduction

Melamine is successfully used in flexible PU foams and other polymers as a halogen-free flame retardant (FR) [1–3]. The melamine itself, upon combusting of a polymer, acts only as a blowing agent, which dilutes flammable volatile products, released upon the decomposition of a polymer at high temperature [4]. To reach a greater effect and owing to synergism of nitrogen and phosphorus, it is often used in combination with ammonium polyphosphate, which acts as charring catalysts in intumescent FR systems. Another type of FR, based on the well-known phosphorus/nitrogen synergism [5], are phosphorous salts of melamine.

Melamine phosphate and melamine polyphosphates of different grades under various names are commercially available. They are mainly used in different thermoplastics, such as polypropylene and polyamide [6, 7]. It is recommended to use melamine phosphate also in thermosets, both unsaturated polyester and epoxies [8, 9]. At the same time, there is not so great experience of using melamine phosphate in cross-linked polyurethanes [10, 11].

In the present study, the efficiency of melamine phosphate as an intumescent FR in a polyurethane coating based on triethanolamine esters of tall oil fatty acids was assessed. The effects of the melamine phosphate content and the weight proportions in its combination with other FRs on the main parameters of the thermal stability and flammability of polyurethane coating were studied. The mentioned parameters of the new coatings were compared to the parameters of similar polyurethane coatings with most often used intumescent FRs – single ammonium polyphosphate, as well as ammonium polyphosphate in combination with melamine.

2. Experimental

2.1. Materials
Melamine phosphate (P > 12%, N > 42%) was purchased from Tianjin Aokatet Chemical Co., Ltd. (China). Ammonium polyphosphate, grade Exolit® AP 422 (P 31.5%, N 14.3%) was obtained from Clariant International Ltd., BU Additives (Switzerland). Melamine 99% (N 66.8%) was purchased from Sigma-Aldrich Chemie GmbH (Germany).

Polyisocyanate Voratec SD 100 (NCO content 31.5%, functionality 2.7) was obtained from Dow Deutschland GmbH (Germany). Triethanolamine esters of tall oil fatty acids (OH number 349 mg KOH/g), synthesised as in [12], were used as a polyol.

2.2. Preparation and testing of polyurethane coating
Melamine phosphate (MP), in mixture with polyol, did not behave as inert filler; therefore, it was preliminary mixed in polyisocyanate. Ammonium polyphosphate (APP) and its mixtures with melamine (MEL) like in [13] were preliminary mixed in polyol. Then toluene (up to 50%) was added to the polyol or polyisocyanate premixed with intumescent additives. Polyurethane coatings were prepared at the molar ratio NCO/OH = 1.1. For the thermal test polyurethane coatings were prepared in the form of free films. For the flammability test, polyurethane coatings, 250-300 μm in thickness, were applied on 100×100×16 mm standard wood (pine) samples.

The consumption of the polyurethane composition upon its applying to both polypropylene plates (free films) and wood samples was controlled by the weight method. All polyurethane coatings were hardened at a temperature of (21 ± 2)°C during 7 days.

TGA/SDTA 851e METTLER TOLEDO was used for thermal gravimetric analysis (TGA) of polyurethane coatings. Sample weight was about 8 mg; heating rate 10°C/min. The test was carried out in the airflow of 20 cm^3/min. FTT Cone Calorimeter (Fire Testing Technology Ltd.) was used for the combustibility test of wood samples with applied coatings. Testing was performed according to ISO 5660 at a heat flux of 35 kW/m^2 like in [13]. All samples were arranged in a horizontal position. Test duration was 30 min.

3. Results and discussion

3.1. TGA of intumescent polyurethane coatings
At first, the effect of the content of each flame retardant separately on the thermal stability of polyurethanes was investigated. It was found that, although the decomposition temperature of MP (350°C) is higher than that of APP (300°C), the type of the flame retardant did not exert any essential effect for the first step of filled polyurethanes' decomposition. With increasing content of both the flame retardants, both the initial decomposition temperature and the temperature of the maximum rate of weight loss for the first steps of decomposition of polyurethanes remained practically on the same level (~230°C and 260°C, respectively). Actually, both those temperatures were determined by the lower initial temperature of destruction of the polyurethane itself.

The effect of the flame retardant's type manifested itself only for the final step of polyurethane decomposition. At an equal content of flame retardants, the char yield for the APP-containing polyurethane was much higher than that for the polyurethane, containing the same amount of MP (Fig. 1). This was because the content of phosphorus in APP was almost 2.6 times higher than in MP. No significant effect of the much higher content of nitrogen in MP (2.9-fold) on the final results of destruction was found.

Then the effect of the weight ratio of flame retardants in the combinations APP/MP and APP/Mel at the same total content of flame retardants (25 wt.%) was investigated. It was exactly this content of flame retardants, at which most of the investigated indices of polyurethanes reached their optimum values. With varying ratio, in the mentioned combinations, no essential difference in the indices, characterising the first step of polyurethane decomposition, was observed. At the same time, by varying the weight ratio from 1:1 to 6:1 in the mentioned combinations of FRs (A/B), the char yield at 600°C (m_{600}) of filled polyurethanes slightly increased (Fig. 2). In this case, it exceeded the value of the char yield of polyurethane, filled with single MP, and approximated the value of the char yield of polyurethane, filled with single APP.

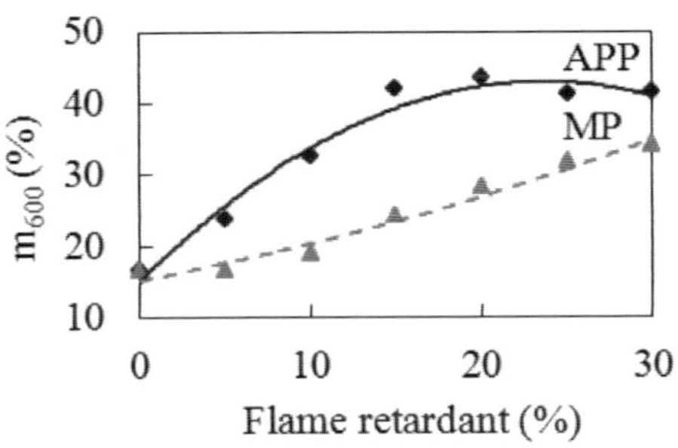

Figure 1. Char yield at 600°C for polyurethanes versus FRs' weight content.

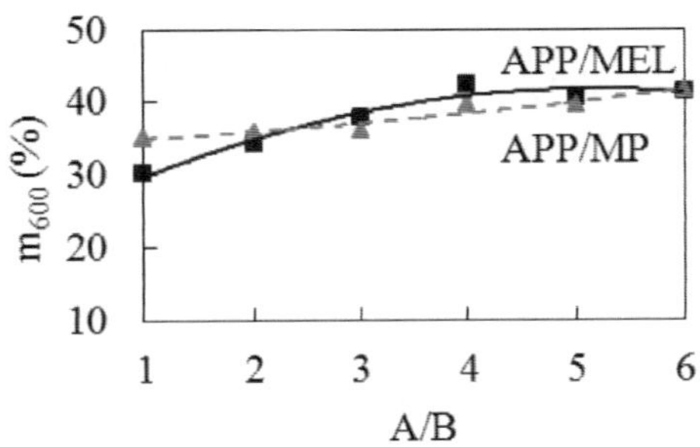

Figure 2. Char yield at 600°C for polyurethanes versus FRs' weight ratio.

3.2. Cone calorimeter study of polyurethane coating

The flammability parameters of wood samples with polyurethane coatings were measured in terms of time to ignition (t_{ig}), heat release rate (HRR), peak heat release rate (PHRR), total heat release (THR), mean mass loss rate (MLR) and total smoke release (TSR) [14]. Additional parameters, which better characterised the flame resistance of materials such as maximum average rate of heat emission (MARHE) and fire growth index FGI = PHRR/t_{ig} [15], were also determined.

Comparing the mentioned parameters of combustion of the samples with polyurethane coatings with the addition of one of the flame retardants, it was found that in terms of the indices, MP as a flame retardant is as efficient as APP. With increasing content of both the flame retardants up to 30 wt.%, the main flammability parameters of polyurethanes varied approximately equally, despite the much lower (2.6-fold) content of phosphorus in MP.

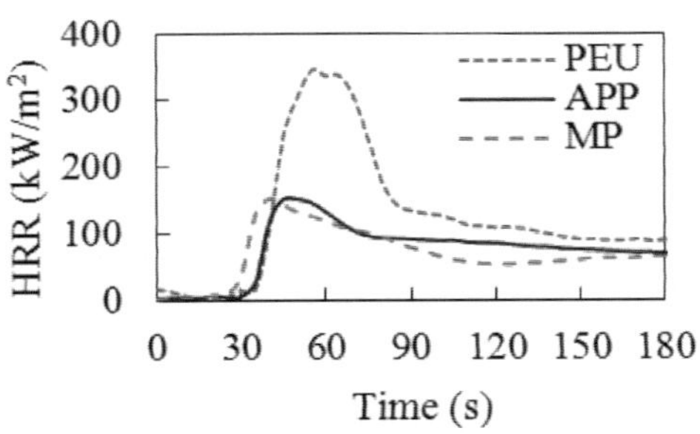

Figure 3. HRR versus time for polyurethane coatings with 25% of FR and without FR (PEU).

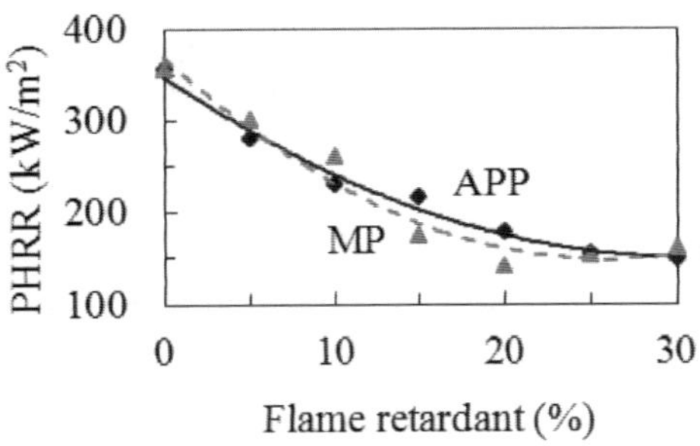

Figure 4. PHRR versus APP and MP weight content for wood samples with coatings.

The parameters, characterising the initial period of burning of the wood samples with polyurethane coatings, varied to a greater extent (Fig. 3). Thus, with increasing content of both MP and APP up to 25%, PHRR for the samples with a polyurethane coating decreased 2.3 times (Fig. 4). The decrease of the FGI of the samples with polyurethane coatings, containing 25% of MP or APP, was still greater, namely, 2.8-fold.

The flammability parameters, characterising the behaviour of samples for the full time of the test, did not decrease so much. Thus, MARHE for the wood samples with a polyurethane coating, with increasing content of FRs up to 25%, decreased 1.8 times (Fig. 5). However, THR and MLR at loading of the same amount of MP or APP decreased only 1.2 and 1.1 times, respectively.

The almost equal efficiency of those two flame retardants was obviously explained by the additional action of another element in the composition of flame retardants during combustion, namely, nitrogen, as well as its synergistic action with phosphorus. The comparatively moderate content of phosphorus in MP, in comparison with APP, was compensated by the much higher content of nitrogen. The total content of nitrogen and phosphorus in MP was also higher (1.2-fold) than in APP.

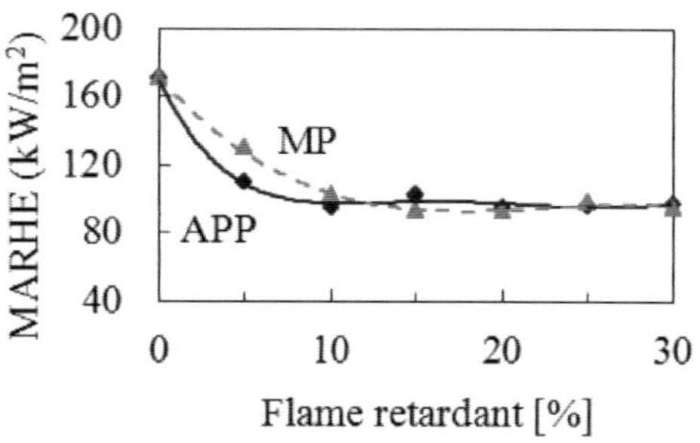

Figure 5. MARHE versus APP and MP weight content for samples with coatings.

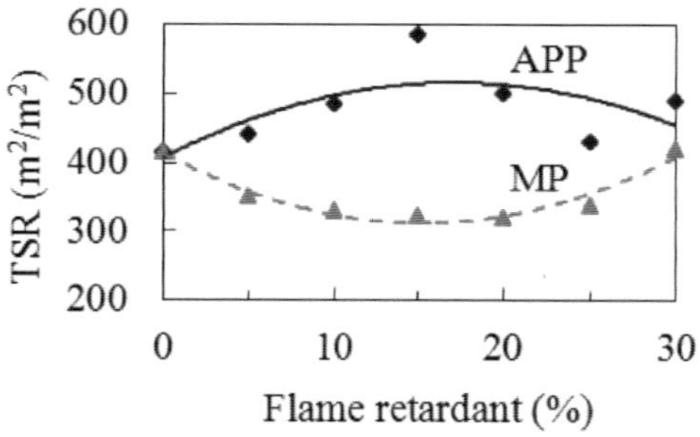

Figure 6. TSR versus APP and MP weight content for wood samples with coatings.

Owing to the high content of nitrogen MP had one distinct advantage. As in the case of adding of melamine (blowing agent) [13], the TSR of samples with polyurethane coatings, containing the additions of MP, were much lower than those for the same ones with additions of APP. With increasing content of MP up to 20%, TSR of the samples decreased sequentially (Fig. 6). With the further increase of the MP content, TSR of the samples with a polyurethane coating increased, probably due to the insufficiently good dispersion of MP in the coating at its high content.

Taking into account the lower smoke release, MP, as a single flame retardant, appeared to be more preferable than single APP. In this case, the content of MP, at which the flammability parameters of polyurethane compositions reached their minimum, was lower (20%) than for APP (25%).

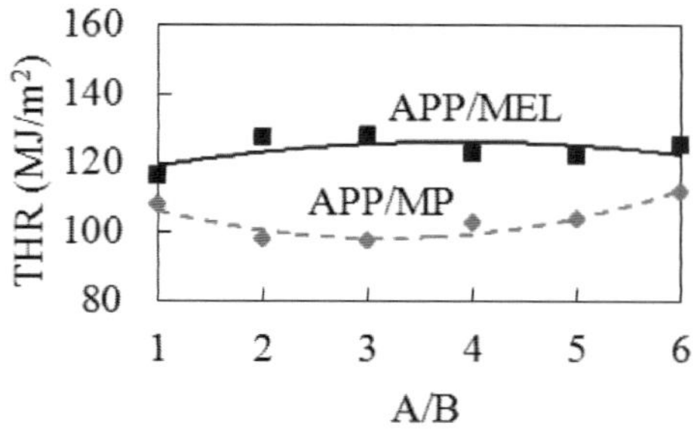

Figure 7. THR versus FRs' weight ratio for wood samples with coatings.

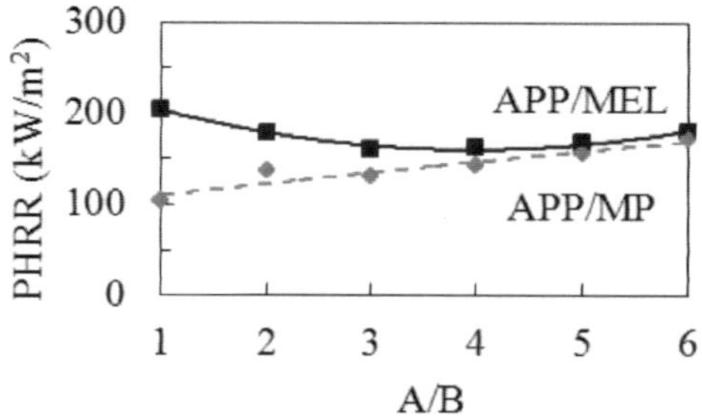

Figure 8. PHRR versus FRs' weight ratio for wood samples with coatings.

Varying the weight ratio in double combinations of FRs, it was found that most of the flammability parameters of samples with coatings, containing the combination APP/MP, were better than the parameters of the samples with coatings, containing the combination APP/MEL. This could have been expected, because, at the same weight ratio of those FRs in the polyurethanes with the addition of the APP/MP mixture, the content of phosphorus is higher than that in the polyurethanes with the addition of the APP/MEL mixture. With increasing weight ratio in the mentioned combinations from 1:1 to 6:1, the phosphorus content in the polyurethanes with the addition of APP/MP increased from 5.4% to 7.2%, but in the polyurethanes with the addition of APP/Mel, the phosphorus content grew from 3.9% to 6.8%. The total content of phosphorus and nitrogen in the polyurethanes with the addition of APP/MEL was only slightly higher than that in polyurethanes with the addition of APP/MP.

Increasing the weight part of APP in combinations of APP/MP and APP/MEL MLR, TSR and MARHE of the samples with polyurethane coatings changed in the same manner, like THR (Fig. 7). In this case, most of the parameters' values did not reach the levels of the parameters of the samples with a polyurethane coating with the same amount of single MP or APP. Only THR and MARHE of samples, containing the combination APP/MP, was lower by 20% than that of the samples with single MP.

Decreasing the weight ratio in the mentioned FRs' combinations, PHRR of the samples with APP/MEL increased, but with APP/MP decreased (Fig. 8). In the range of the ratios from 3:1 to 1:1, PHRR of the samples, containing the combination APP/MP, was lower by 10-20% than for samples, containing single MP. In this case, these values of THR, MARHE and PHRR were the best of all the earlier obtained results of testing the samples with polyurethane coatings, containing single intumescent FRs or their various combinations [13].

4. Conclusions

Thus, MP as single intumescent FR, decreases most of the flammability parameters of the polyurethane coatings based on tall oil fatty acids, like single APP. It was found that the maximum effects on flammability parameters took place at loading 20% of MP into the polyurethane coating. In contrast to APP, the addition of MP reduces also the smoke release of the samples. Using MP in combination with APP (up to 50% from the total amount of FRs) it is possible to essentially reduce the flammability parameters of polyurethane coatings, such as PHRR, THR and MARHE.

Acknowledgement

The present study was supported by the State Research Program 4.1 "Forest and earth entrails resources: research and sustainable utilization – new products and technologies" (ResProd).

References

[1] Levchik S V and Weil E D 2004 *Polym. Int.* **53** 1585
[2] Weil E D and Levchik S V 2004 *J. Fire Sc.* **22** 183
[3] Wang J Q and Chow W K 2005 *J. Appl. Polym. Sc.* **97** 366
[4] Lowden L A and Hull T R 2013 *Fire Sc. Rev.* **2** 4
[5] Liang S, Neisius N M and Gaan S 2013 *Prog. Org. Coat.* **76** 1642
[6] Zhou S, Song L, Wang Z, Hu Y and Xing W 2008 *Polym. Degrad. Stab.* **93** 1799
[7] Kiliaris P, Papaspyrides C D, Xalter R and Pfaendner R 2012 *Polym. Degrad. Stab.* **97** 1215
[8] Chen W Y, Wang Y Z and Chang F C 2004 *J. Polym. Res.* **11** 109
[9] Weil E D 2011 *J. Fire Sc.* **29** 259
[10] Hassan M A, Kozlowski R and Obidzinski B 2008 *J. Appl. Polym. Sc.***110** 83
[11] Costache M C, Kanugh E M, Wilkie C A and Sorathia U 2006 *J. Fire Sc.* **24** 433
[12] Yakushin V, Stirna U, Bikovens O, Misane M, Sevastyanova I and Vilsone D 2013 *Mater. Sc. (Medžiagotyra)* **19** 390
[13] Yakushin V, Sevastyanova I, Vilsone D and Kirpluks M 2015 *Mater. Sc. (Medžiagotyra)* **21** 226

IOP Conf. Series: Materials Science and Engineering **111** (2016) 012002 doi:10.1088/1757-899X/111/1/012002

[14] Babrauskas V and Parker W J 1987 *Fire Mater.* **11** 31
[15] Schartel B and Hull T R 2007 *Fire Mater.* **31** 327

Some Exploitation Poperties of Wood Plastic Composites (WPC) Based on High Density Polyethylene (HDPE) and Plywood Poduction Waste

J Kajaks[1] A Kolbins[2] and K Kalnins[2]

[1]Assoc.professor, Institute of Polymer Materials,Faculty of Materials Science and Applied Chemistry, Riga Technical University, Riga, LV.

[2] Master of. science.Institute of Polymer Materials,Faculty of Materials Science and Applied Chemistry, Riga Technical University, Riga, LV.

E mail:kajaks@ktf.rtu.lv

Address. Faculty of Materials Science and Applied Chemistry,Riga Technical University, P. Valdena str. 3/7, Riga, LV1048, Latvia

Abstract. A lot of researchers are connected with partially green, biocomposites producing and investigations. Various of polymer matrices, mainly polyolefines, reinforcement-natural fibres and their combinations are used. Our studies are focused on high density polyethylene based biocomposites, containing investigation results of plywood production by-product sanding dust (PSD) exploitation properties (flexural strength and modulus, impact strength, microhardness, and water uptake). Also, the fluidity of composite melts was noted, but the fracture mechanism of the composites tryed to clear up with SEM studies. For wood plastic composites (WPC) modification coupling agents maleatedpolypropylene (MAPP) and polyvinylbutyral (PVB) were used.

Introduction

During the last 20-30 years many researchers have paid attention to the studies of exploitation properties of wood polyolefine composites(WPC) [1-7]. The most useful from the wide selection of polyolefines are high density polyethylene (HDPE) or polypropylene, but as reinforcement of WPC different waste materials comprising lignocellulose fibres often are used. Such materials also are a lot of waste which arises, for example, in plywood production industry [8]. In this work for reinforcing of HDPE we approbated one of the by-product-plywood sanding dust (PSD), which contain fractions of wood fibres with different length (l). The main fraction (68%) are fibres with length of l=0.25-0.5 mm. Our previous studies [8] showed that plywood sanding dust (PSD)is promising filler for PP reinforcing. Additions of the PSD up to 40 wt.% increase tensile and flexural modulus, but decrease deformation ability of PP matrix, impact strength, water resistance and fluidity of composite melts. Modification of composites with interfacial modifier maleated polypropylene (MAPP up to 5-7 wt.%) considerably improve all the above-mentioned properties. Also, very popular matrix for preparing WPC also is different types of recycled HDPE [4,9].In the work[9] was shown that for composites based on virgin HDPE, recycled rHDPE and grinded coniferous tree (pine) shavings (length of wood fibres up to 0.5 mm) properties considerably increase: flexural strength up to 2, but modulus up to 3 times to compare with pure polymer matrices. Maximum bending deformation and impact strength values decrease till 1.2-1.8% and 6.28, 4.61 kJ/m^2 for rHDPE and virgin HDPE respectively. All WPC practically do not swell in water and total amount of absorpted water after 250 h water uptake is not more than 4-8%. Noteworthy, microhardness of the samples increase (2-3 times) from 70 MPa for rHDPE up to 150-200 MPa for composites with 50-60 wt.% of the wood fibres. It means, that it is possible to use these systems for production of WPC materiāls and different products from WPC.The goal of this study was to determineplywood production by-product plywood sanding dust (PSD) influenceon some virgin HDPE composites physical-mechanical and other exploitation properties.

Materials and methods

As a polymer matrix virgin HDPE type Liten PND 33-300, (MFI=15.58g/10min.) was used.Filler wasplywood production by-product birch wood plywood sanding dust (PSD)and content in the composites were(30, 40, 50, 60 wt.%). As an interfacial modifier maleated polypropylene type Licocene PP MA-7452 (1,3,5wt.%) and polyvinylbutyralgrade B 60HDE (content 5,10 wt.%) were used. Composites were prepared by mixing on a two-roll mill, then cooled, granulated and pressed in 1 mm thick sheets for microhardness measurements by Vikers M-41.Flexural and impact strength tests were done for standard specimens (EN ISO 178 and ASTM D 256M respectively), bars produced by injection moulding. Melt fluidity (melt flow index-MFI) was estimated by standard ASTM D 1238. Water resistance measurements were done according to standard ASTM D 570-88. Microstructure of the fracture surface of specimens were examined with scanning electronic microscope Vega Tescan 5136 MM. Au-sputtering was used.

Results and discussion

First of all, for clearing up of the optimal content of plywood sanding dust (PSD) in HDPE some physical-mechanical test measurements for the composites with different amount of PSD fibreswere done. Results of these experiments are given in Table 1.

Table 1.

Physical-mechanical test results for the composites based on high density polyethylene (HDPE) and plywood sanding dust (PSD)

Parameters	Concentration of PSD, wt.%				
	0	30	40	50	60
Flexural strength, MPa	17.20	25.46	26.74	27.06	26.72
Flexural modulus, MPa	449.3	1222.5	1558.8	1863.0	2218.9
Maximum of bending deformation, mm	12.13	8.26	5.31	3.33	2.50
Impact strength, kJ/m^2	26.03	13.31	7.50	5.78	4.90
Microhardness, MPa	58.57	75.05	92.56	112.96	115.74

From presented results (see Table 1) we can conclude, that all investigated properties improve with increase of PSD content in composites up to 50 wt.%. Extension of the filler content till 60 wt.%give only small increases of microhardness and flexural modulus, but decreases flexural and impact strength. If it is necessary to process gained composites in different products, then very important parameter is fluidity of the composite melts which is able to evaluate by melt flow index (MFI) measurements.PSD loading in HDPE considerably (more than 20 times) diminish polymer matrix fluidity from 15.6 g/10min. for virgin HDPE till 0.62 g/10 min. for system with 60 wt.%. Such a dramatic decrease of MFI values can make worse possibilities to process these materials in different products, especially by injection molding method. Bearing in mind previous physical-mechanical experimental results and fluidity of the composite melts, as the optimal content of PSD in a composite was chosen 50 wt.%. For improving WPC exploitation properties, frequently coupling agents like maleated polypropylene (MAPP) are used [1,6,8]. Additions of the MAPP and PVB to HDPE+50wt.% PSDin the aggregate improve exploitation properties of these composites (see Table 2).

Table2

Coupling agents influence on exploitation properties of composites based on HDPE and 50 wt.% PSD.

Parameters	Concentration of MAPP, wt.%				
	0	1	3	5	5 wt.% PVB
Flexural strength, MPa	27.06	26.92	28.98	28.05	27.52
Flexural modulus, MPa	1863.0	1607.5	1633.5	1696.6	2067.8
Maximum of bending deformation, mm	3.33	4.17	4.10	3.94	2.46
Impact strength, kJ/m^2	5.78	5.88	5.88	5.43	4.84
Microhardness, MPa	112.9	107.5	155.3	121.8	112.1

Flexural strength increase up to 28.98 MPa (3 wt.% MAPP) and modulus up to 2067.5 MPa (5 wt.% PVB). Impact strength properties of composites was optimal at 3 wt.% MAPPconcentration. At the same time, melts of these composites maintain rather good fluidity (MFI values are 2.65 and 2.4 g/10 min.for composites with 50 wt.% PSD+3 wt.% MAPP and 50 wt.%.PSD+5 wt.%. PVB respectively). The greatest microhardness showscomposites containing 50 wt.% PSD (112.96 MPa) and 50 wt.% PSD + 3wt. % MAPP (155.26 MPa). It is well known fact [1,2], that coupling agents like MAPPpromote better distribution of lignocellulose fibres in polymer matrix, intensify interfacial interaction between polar fibres and nonpolar polyolefine matrix. Obviously, these phenomenons take place also in investigated systems, for example microhardness increase up to 155.3 MPa. From this point of view and bearing in mind high cost of MAPP, as optimal concentration of the MAPP in composite we chose 3 wt. %. Besides, at the higher concentrations of the MAPP can be observed its plasticizing influence on WPC properties [1,8]. Our presented investigations also showed positive influence of the MAPP on WPC melts fluidity. MFI values increase at the presence of MAPP from 2.15 g/10min. for nonmodified system up to 2.65 and 3.4 g/10min.for composites with 3 and 5 wt.% MAPP correspondingly. Additions of interfacial modifier (3-5 wt.% MAPP) also promotethe increase of water resistance (see Fig.1).Water absorption after 28 days water uptake decreases from 10% for unmodified composite (curve 1) till 5.7-6% formodified with 3 and 5 wt.% MAPP systems (curves 2,3) that indicates positive action of the MAPP, strengthens interfacial adhesion between components. Samples thickness swelling also is small at the presence of the MAPP and fluctuate in the range of 1-2%.

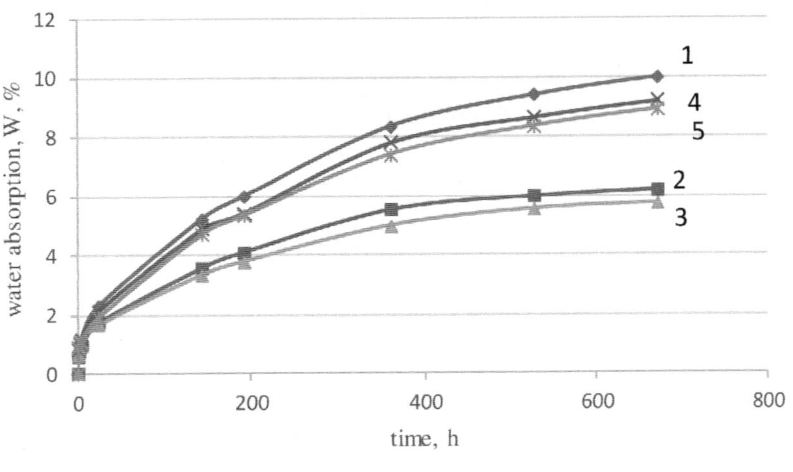

Figure 1. Water absorption kinetics of high density polyethylene (HDPE) composites. 1- HDPE+50 wt.% PSD, 2-HDPE+50wt.% PSD + 3 wt.% MAPP, 3- HDPE+50 wt.% PSD+5 wt.% MAPP, 4- HDPE+50 wt.% PSD +5 wt.% PVB, 5- HDPE+50 wt.% PSD+10 wt.% PVB.

IOP Conf. Series: Materials Science and Engineering **111** (2016) 012003 doi:10.1088/1757-899X/111/1/012003

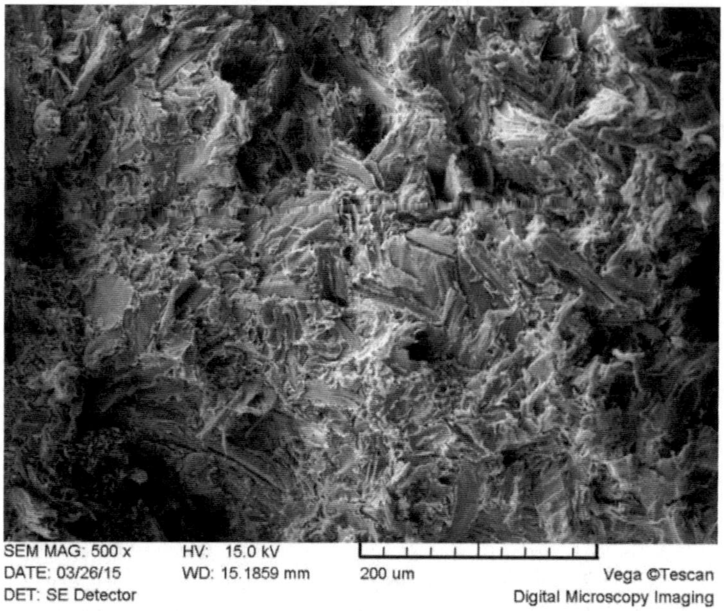

a

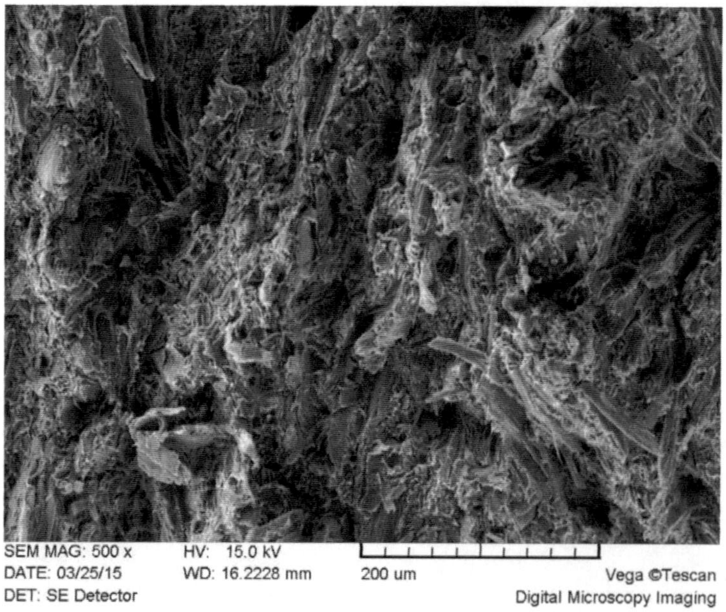

b

Fig.2 a,b. Scanning electron microscopy pictures (SEM) of the fracture surface of specimens made from different composites: a-HDPE+50 wt.% PSD, b-HDPE+50wt.% PSD + 3 wt.% MAPP.

In order to clear up coupling agent influence on the strengthening effect of composites and fillers distribution degree in polymer matrix, scanning electron microscopy (SEM) investigations were used (see Fig.2 a,b). Loking over SEM pictures, replicas of a fracture surface of unmodified material (Fig.2 a) partially or completaly uncovered particles are observed. That indicates poor interfacial interaction between wood fillers and high density polyethylene. Besides, distribution of the fibres in polymer is irregular. Coupling agent MAPP additives strengthen border between wood fibres and HDPE that provides evidence, that interface surface only partially is broken (see Fig.2 b) and the fine particles in polymer matrix hold well. SEM studies once more confirmed the fact of strengthening the interfacial

IOP Conf. Series: Materials Science and Engineering **111** (2016) 012003 doi:10.1088/1757-899X/111/1/012003

adhesion between wood fillers and HDPE matrix as a result of the action of the MAPP, that evidently givesi ncrease in exploitation properties of the wood (PSD)high density polyethylene composites.

Conclusion

1.Wood (PSD) high density polyethylene composites physical mechanical properties studies showed that loading of PSD considerably increase flexural strength, modulus and microhardness of WPC and optimal concentration of plywood sanding dust content in HDPE could be 50 wt.%.

2.Modification of WPC with MAPP and PVB showed that coupling agents improve actually all exploitation properties of HDPE+50 wt.% PSD composites compared to unmodified systems and optimal content of the MAPP is 3-5 wt.%. Particulary, remarkable decrease of water absorption and the best water resistance was demonstrated by composites with 5 wt.% MAPP.

3. SEM studies confirmed the MAPP strengthening effects of the interfacial adhesion between wood fillers and high density polyethylene matrix andplywood production by-product sanding dust (PSD) is useful for preparing wood plastic composites based on a high density polyethylene.

References

[1]Polyolefin Composites Ed. by Domasius Nwabunma and Thein Kun 2007 3M Company, Wiley-IntersciencеA. John Wiley and Sons INC Publications, pp.3-82, 87-123, 150-201.
[2]Kumar V et al 2011 A Review./ Reviewers in Chem. Engineering, **27**(5-6) 252
[3]Ogah A, Alfukwa J 2014 J.of Reinf. Plastics and Comp. **33** (1) 37
[4[Wambua P et al 2012 Research J. of Appl. Sc. Engineering and Technol. **4** (7) 703
[5]Agyarwall P et al 2013 J. of Reinf. Plastics and Comp. **32**(22) 1422
[6] Kord B 2012 World Appl. Sc. Journal **17**(2) 168-171.
[7]Kaboorani A and Englund K 2010 J. of Comp. Mat.**45** (13) 1423
[8]Kajaks J et al 2014 Central European J. of Engineering. **4** 385
[9]Kajaks J et al 2015 Materials Science (Medziagotyra) **21** (3) 396-399

IOP Conf. Series: Materials Science and Engineering **111** (2016) 012004 doi:10.1088/1757-899X/111/1/012004

Functional properties of composite material from recycled tires and polyurethane binder in water medium

R.Plesuma, L.Malers

Institute of Polymer Materials, Faculty of Materials Science and Applied Chemistry, Riga Technical University, P.Valdena 3/7,Riga, LV-1048,Latvia.

E-mail: laimis@ktf.rtu.lv

Abstract. The present research is as a continuation of the authors' previous research of composite material and practical application of composite material largely connected with water. The aim of present study was to esablish certain functional properties of the material in water medium. Water permeability , absorption and swelling of the composite material after being exposed to water for certain period were determined. Water absorption , permeability and swelling of the composite material showed close correlation with polymer reactivity. Molding pressure , temperature and the distribution of rubber particle sizes also demonstrate a direct influence on the water absorption and permeability of the composite material. The obtained results are useful for the practical application of selected composite material with desirable and predictable functional properties.

1.Introduction

Production of composite material is considered as one of the most perspective directions of scrap tires reuse in combination with polymer binder [1-3]. In our previous studies optimization of composition and technology of the production of composite material from mechanically grinded scrap tires and polyurethane type binder were carried out [4]. It was also established, that mechanical properties of the composite material are strongly dependent not only on the composition of material and the reactivity of polymer binder, but also on technological parameters (molding pressure, temperature) and environmental conditions (relative air humidity RH) [5-8].

The present study is a continuation of the authors's previous research of composite material and practical application of composite material largely connected with water. The aim of the present study was to estabilish certain functional properties of the material in water medium.

2. Materials and investigation methods.

In this work unfractionated und fractioned scrap rubber crumb was used in order to produce samples of composite material. The particle size distribution of unfractionated rubber crumb is determined by modified Sieve method based on LVS EN 933-1. Polyurethane binder with differing content of reactive isocyanate -*NCO* groups (1.8%, 2.4% and 7.4%) was used. Different quantities of binder content (8,13,18 and 23 wt. %) was introduced based on the considerations of the previous investigations [4,5]. Specimens of composite material were prepared by mechanical mixing of scrap rubber crumb with polymer binder at definite formation and hardening temperature T, molding pressure P, relative air humidity RH (40-45%) and hardening time (24h).Water permeability (introduced method), absorption and swelling of the composite material after being exposed to water for a specific period were determined according to LVS EN 12087.

3. Results and discussion.

The degree of water absorption and swelling of composite material samples in water was determined after 7 days of being exposed to water. Water absorption was evaluated not only in connection with the temperature of composite material formation but also with reactivity of polymer binder (figures 1 and 2).

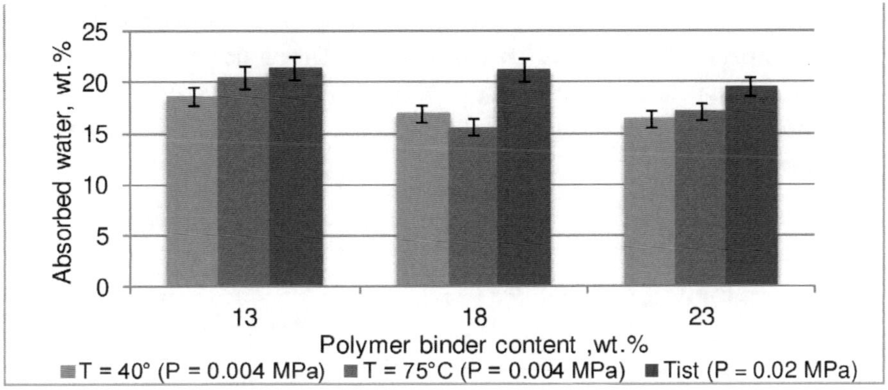

Figure 1. Variation of absorbed water in composite material with polymer binder content and conditions of samples producing (- NCO 2.4%).

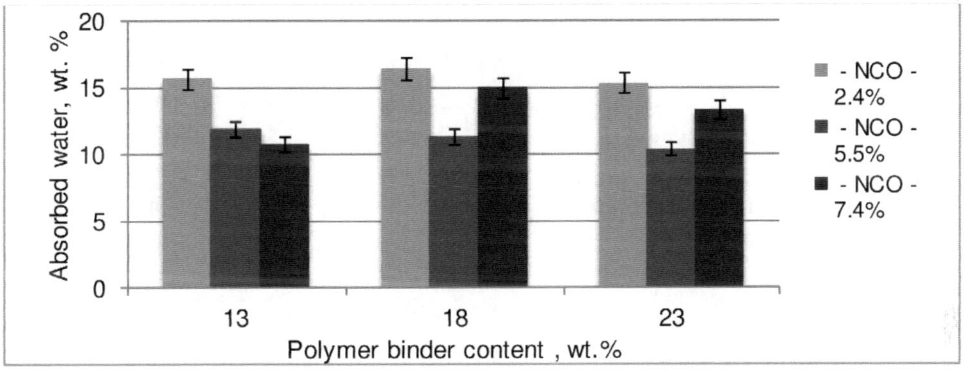

Figure 2. Variation of absorbed water in composite material with polymer binder content and reactivity (-NCO content).P= 0.004 MPa, T=18-20^{0}C.

It was observed , that after a certain exposition time (168 h) in water , the quantity of absorbed water is closely connected with the polymer binder content. A smaller amount of polymer in composite material provokes absorption of water due to the increase of the voids in material ascertained in previous research [9]. The results show that both the temperature and the molding pressure as a technological parameters have definite influence on absorbed water in composite material (figure 1). It can probably be explained by a higher crosslinking degree of the polymer binder at higher formation temperatures although a lower molding pressure was used.

The influence of polymer binder reactivity on water absorption was also ascertained (figure 2).The results show that variation in absorbed water in composite material with smaller polymer binder reactivity present more even changes with polymer binder content than with larger binder reactivity. As it was observed, an increase in the degree of polymer binder crosslinking with increase of polymer binder reactivity promotes a decrease in the absorbtion capacity of the composite material.

IOP Conf. Series: Materials Science and Engineering **111** (2016) 012004 doi:10.1088/1757-899X/111/1/012004

The correlation between the swelling degree of composite material, the polymer binder content and reactivity of polymer used in composite material as well as the formation temperature has been studied (figures 3 and 4). It was observed, that not only the reactivity of the selected polymer binder, but also the potential of obtaining a more compact structure by increasing the binder functionality and content in the material is responsible for the suggested correlations.
A direct influence of the formation temperature on the swelling degree was demonstarted (figure 3). The swelling degree is higher for samples produced at T=40°C. It can be explained with a direct influence of higher formation temperature on a higher crosslinking degree of polymer and therefore the swelling properties of the material. It is also closely correlate with the results of the previously conducted studies regarding the investigation of Shore hardness of similar samples produced at different temperatures [8].

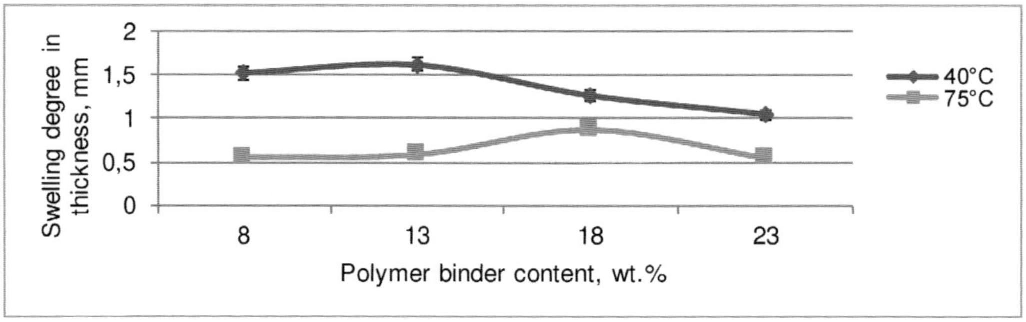

Figure 3. Variation in the swelling degree in thickness with a polymer binder (-NCO 2.4%) content at different formation temperatures. P= 0.004 MPa .

The swelling degree in mass of composite material was investigted useing a polymer binder with a different reactivity (figure 4).

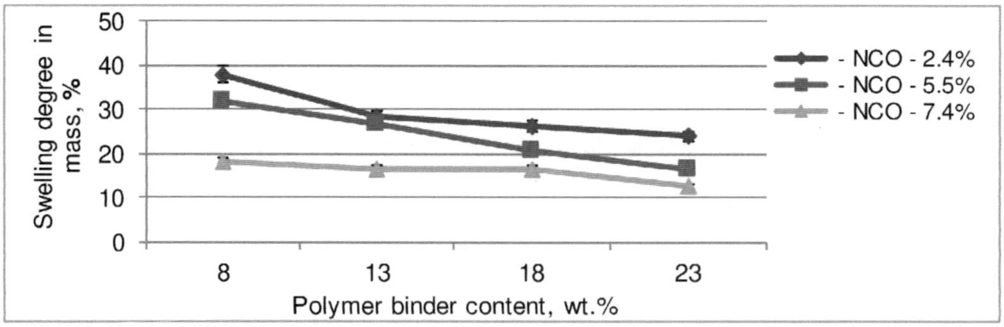

Figure 4. Variation in the swelling degree in the mass with polymer binder reactivity and content.

P= 0.004 MPa, T=18-20°C.

The correlation was demonstrated between the swelling rate in the mass and polymer binder reactivity and content in composite material. The obtained results can be explained with accumulation of voids attainable for water in composite material with decreasing polymer binder content. The less the reactivity of polymer binder and therefore the crosslinking degree, the higher the swelling degree of the material was observed (figure 4).

Water permeability (permeability time) of composite material was investigated with a method according to which 300 ml of water penetrate 15.2 cm^2 material sample. A correlation between the permeability time, rate of polydispersity of rubber crumb and the polymer binder content is observed (table1).

Table 1. Dependence of the composite material water permeability time from polydispersity of rubber crumb and polymer binder content . P= 0.004 MPa, T=18-20^{0}C.

Polymer binder content, wt.% (-CNO 2.4%)	Permeability time, sec. (rubber crumb with less polydispersity)	Permeability time, sec. (rubber crumb with higher polydispersity)
13	10.0	7.6
18	10.8	8.2
23	12.4	9.0

The results show that permeability of samples decreases with an increase of polydispersity of the rubber crumb and polymer binder content.This can possibly be explained by the reducion of voids and therefore the formation of a more compact structure of the composite material. It was confirmed especially if fractionated rubber crumb was introduced beforehand in the composite material (figure 5).

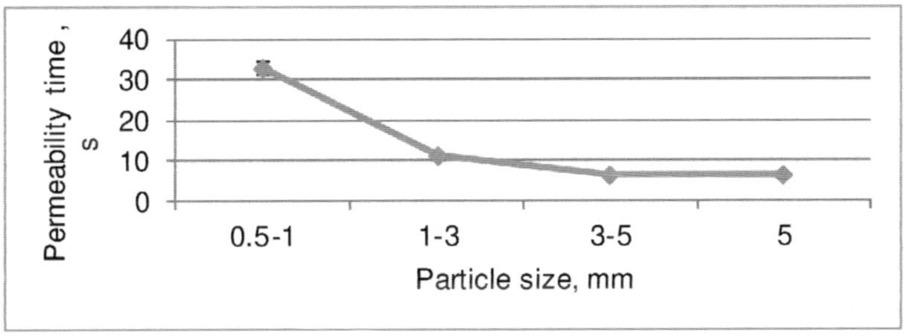

Figure 5. Variation in water permeability time with rubber particle size (polymer binder content 18 wt.%.with -NCO 2.4%). P= 0.004 MPa, T=18-20^{0}C.

A decrease in water permeability was observed with an increase of the molding pressure, probably due to the formation of a denser structure of the composite material (figure 6).

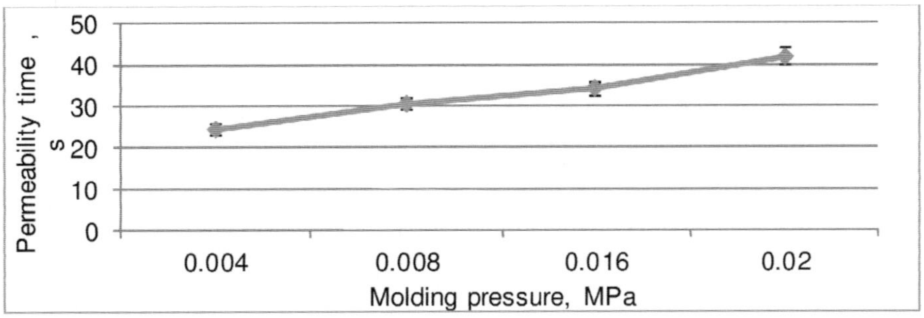

Figure 6. Variation in water permeability time with molding pressure (polymer binder content 18 wt.%.with -NCO 2.4%). P= 0.004 MPa, T=18-20^{0}C.

Therefore the assumption, that the composite material crosslinking degree and the presence of possible voids must be regarded as important factors having influence on water permeability, was confirmed.

4. Conclusions.

1.Water absorption and swelling of the composite material showed a close correlation with the reactivity of polymer and thus underline the importance of the crosslinking degree of the polymer binder.
2.The molding pressure and the rubber particle size distribution also demonstrate a remarkable influence on the water absorption and permeability of the composite material, but temperature as a technological factor has a significant influence on the swelling rate of the material.

.

5. References

[1] Mark J E, Erman B, Erich R *The Science and Technology of Rubber* 2006 ,3 d.Ed., Elsvier Inc., USA.
[2] Moiliner F,Lardner K,Dunster A *Waste & Resources action programme* 2007 ISBN:1-84405-351-2
[3] Ayrilmis N, Buyuksari U, Avci E *Utilization of Waste Tire Rubber in Manufactoring of Particle Board, Materials and Manufacturing Process* 2009 **24:6** 688-692
[4] Malers L, Plesuma R, Locmele L *Mechanics of Composite Materials* 2009 T**45** Nr.1 145-150
[5] Malers L,Plesuma R, Locmele L,Megne A,Mateusa-Krukle I , Kalnins M *Abstracts of Baltic Polymer Simposium - 2010, Kaunas University of Technology* 2010 55
[6] Malers L.,Plesuma R ,Locmele L *Scientific Journal of Riga Technical University, Material Science and Applied Chemistr* 2011 **23** 103-106
[7] Plesuma R , Megne A, Mateusa-Krukle I,Malers L *Progress in Rubber, Plastics and Recycling Technology*, 2013 **29** Nr.3 177-188
[8] Plesuma R, Malers L *European Central Journal of Engineering* 2013 **3** 492-496
[9] Plesuma R , Megne A, Mateusa-Krukle I, Malers L *Materials of 17. International Conference, Mechanics of Composite Materials 2012, Riga, Book of Abstracts* 172**.**

IOP Conf. Series: Materials Science and Engineering 111 (2016) 012005 doi:10.1088/1757-899X/111/1/012005

Damage Identification in Polymer Composite Beams Based on Spatial Continuous Wavelet Transform

R Janeliukstis[1], S Rucevskis[1], M Wesolowski[2], A Kovalovs[1] and A Chate[1]
[1]Institute of Materials and Structures, Riga Technical University, Riga, Latvia
[2]Department of Structural Mechanics, Faculty of Civil Engineering, Environment and Geodesy, Koszalin University of Technology, Koszalin, Poland

e - mail: Rims.Janeliukstis_1@rtu.lv

Abstract. In present paper, the damage identification in two polymer composite beams is shown with two methods – spatial continuous wavelet transform (CWT) and mode shape curvature squares (MSCS) using experimental data from operational deflection shapes (ODS). Damage was introduced via low velocity impact drop tower. Statistical hypothesis approach was used to calculate standardized damage index (SDI) that served as a damage indicator for both methods. In order to truncate smaller magnitude SDI peaks, a threshold of 1.28, corresponding to the confidence level of 90 %, was applied. Damage estimate reliability (DER) vs wavelet scale test was performed in order to quantify the reliability of damage detection in terms of percentage at each scale parameter. Overall, 78 different wavelet functions were tested. Results indicate that both methods are capable to locate the area of damage.

1. Introduction

Composite materials possess superior mechanical properties as compared to the conventional materials. On the other hand, these properties of composites may be heavily deteriorated by the presence of such failure modes as matrix cracking, fibre pull-out, fibre fracture, fibre-matrix debonding and delamination between plies. Damage of any structure may cause failure, leading to tragic consequences, but it is especially true for engineering and civil structures. As a result, non-destructive structural health monitoring (SHM) methods have become an important research area in civil, mechanical and aerospace communities.

Over the years, researchers have proposed numerous vibration-based damage detection methods, that are based on the fact, that changes in dynamic characteristics, namely, natural frequencies, operation deflection shapes (ODS) and damping, are directly related to changes in stiffness of the structure. However, the major drawback of those methods is a need for the data of healthy structure [1], which sometimes can be difficult or even impossible to obtain. Possible solutions to this issue are employment of Gapped Smoothing Techniques to generate a smoothed surface of mode shape curvature, thus simulating a healthy state of the structure or using wavelet transform (WT) technique. WT technique, as we know it, originated in the 1990`s and was mainly used for signal singularity detection, signal denoising and image compression [2]. As an analyzing technique for damage detection, WT was first used in 1994 on vibration data by Newland [3, 4]. WT was successfully applied for damage detection in composite materials [5-8]. Katunin et al. [5-7] has been using 2D WT in numerical simulations of carbon-fibre reinforced polymer laminate rotor blades with nonlinear geometry containing small notches, honey-comb composite sandwich structures, 10-layered cross-ply epoxy laminated composite plates, reinforced with E-glass cloth, subjected to low-velocity impact damage. Gallego et al. [8] was studying 2D WT ability to detect structural defects as delamination in carbon-fibre reinforced polymer plates.

In this paper, a method based on WT technique for damage identification in beam structure is described in terms of damage estimate reliability (DER) over dilation parameter values of 78 different wavelet functions. Out of these functions only best are chosen and respective results are compared to the well-known mode shape curvature squares (MSCS) method. Effectiveness and robustness of the proposed algorithms are demonstrated experimentally on two carbon/epoxy polymer composite beams subjected to single low-velocity impact damage.

2. Damage detection algorithms

2.1 Continuous wavelet transform

WT is a mathematical method to transform the original signal into a different domain where additional data analysis becomes possible, therefore damage-affected signal portion is revealed. WT employs *wavelets*, which are special functions, containing some oscillations, such that the mean of wavelet function is 0. Although WT is generally used to analyze signals in time domain, in spatial analysis time is replaced with a coordinate, say, x, giving rise to spatial signal $f(x)$ [2].

Continuous wavelet transform (CWT) is given by [9, 10]

$$W_{s,b} = \int_{-\infty}^{\infty} f(x) \cdot \frac{1}{\sqrt{|s|}} \cdot \psi^{*}\left(\frac{x-b}{s}\right) dx = \int_{-\infty}^{\infty} f(x) \cdot \psi_{s,b}{}^{*}(x) dx \quad (1),$$

where asterisk denotes complex conjugation and $\psi_{s,b}(x)$ is a set of wavelet family functions, derived from a *mother wavelet function* $\psi(x)$ by *translating* (parameter b) and *dilating* (parameter s) the $\psi(x)$ [9, 10]. Parameter s is a real and positive number. If $0 < s < 1$, the function is expanded, if $s > 1$, it is compressed.

Equation (1) is used to calculate CWT coefficients, that are extremely sensitive to any discontinuities and singularities, present in the signal $f(x)$, therefore location of damage due to a sudden loss of stiffness can be detected in those ODS, that yield large amplitude wavelet coefficients.

Damage index (DI) for ODS is depicted as follows (recall equation (1))

$$DI_{i\ CWT}^{n} = W_{i\ s,b}^{n} = \int_{L} w_i^n \cdot \psi_{s,b}(x) dx \quad (2),$$

where $DI_i^n{}_{CWT}$ is a DI for each mode, L is a length of the beam, w_n is transverse displacement of the structure, n is a mode number, i is number of grid point in x direction. The following wavelet families were analysed:

complex Gauss wavelets of orders 1-5, *complex Morlet* wavelets of orders 1-4, *Shannon* wavelets of orders 1-5, *Daubechies* wavelets of orders 2-10, *symlet* wavelets of orders 2-8, *coiflet* wavelets of orders 1-5, *Gauss* wavelets of orders 1-8, *biorthogonal* wavelets, *reverse biorthogonal* wavelets, *Haar*, *Meyer*, *discrete Meyer*, *Morlet* and *Mexican hat* wavelets.

In practice, however, experimentally measured ODS are inevitably corrupted by measurement noise, which can lead to false peaks in DI profiles. These peaks could be mistakenly interpreted as damage or they could mask the peaks induced by real damage. Therefore the results for all ODS are summarized, giving rise to summarized DI, which is defined as the average summation of damage indices for all ODS N, normalized with respect to the largest value of each mode

$$DI_i = \frac{1}{N} \cdot \sum_{n=1}^{N} \frac{DI_i^n}{DI_{i\ ,\max}^n} \quad (3).$$

According to [11], the damage indices, determined for each element, are then standardized, according to the concept of statistical hypothesis testing – element i of the structure is classified as either healthy (null hypothesis H_0) or damaged (alternate hypothesis H_1). In order to test the hypothesis, the DI, given in equation (3), is standardized

$$SDI_i = \frac{DI_i - \mu_{DI}}{\sigma_{DI}} \quad (4),$$

where μ_{DI} and σ_{DI} are mean value and standard deviation of damage indices in equation (3), respectively. The decision for the localization of damage is established, based on a pre-assigned classification criterion: choose H_0 if $SDI_i < C_r$ or choose H_1 if $SDI_i \geq C_r$, where C_r is a threshold value. Typical values of C_r, widely used in literature, are 1.28, 2 and 3 for 90 %, 95 % and 99 % confidence levels for the presence of damage.

To quantify the reliability of wavelets to identify damage location, a new parameter, called „*damage estimate reliability*" (DER) is introduced and calculated as follows:

- The whole interval along the axis of the beam (x axis) is split into 3 parts – (a) before damage 0 mm < x < 150 (320) mm, (b) damage 150 (320) mm < x < 200 (370) mm and (c) after damage 200 (370) mm < x < 300 (550) mm for Beam 2 (Beam 1).
- In each of these parts all values of SDI from equation (4) of a respective wavelet are summed up and divided by the number of data points in this particular interval, giving average amplitude ($\overline{SDI_i}$) in a respective part.
- DER is equal to average SDI in zone of damage (part b), divided by the average SDI in all parts combined. It is expressed in percentage as:

$$DER_i = \frac{\overline{SDI_i}(b)}{\overline{SDI_i}(a) + \overline{SDI_i}(b) + \overline{SDI_i}(c)} \times 100\% \quad (5).$$

2.2 Mode shape curvature squares

Most of the mode shape curvature (MSC) damage detection methods require the baseline data of the healthy structure for inspection of the change in the modal parameters due to damage. Often these baseline modal parameters are not available.

In this paper, an interpolation technique with a Fourier series approximation is applied on a MSC data of the damaged structure, generating smooth MSC surfaces that are estimates of the healthy structure. The Fourier series is a sum of sine and cosine functions that describes a periodic signal:

$$\kappa(x) = a_0 + \sum_{i=1}^{n} a_i \cdot \cos(n\omega x) + b_i \cdot \sin(n\omega x) \quad (6),$$

where a_0 models a constant (intercept) term in the data and is associated with the $i = 0$ cosine term, ω is the fundamental frequency of the signal, n is the number of terms (harmonics) in the series [12]. In this study, MSC data was approximated with Fourier series functions of orders 2-8. The obtained coefficient values a_0, a_i, b_i and ω were used to reconstruct the approximation of MSC data using equation (6). The DI is defined as the absolute difference between squares of the measured curvature of the damaged structure and reconstruction of MSC approximation with Fourier series representing the healthy structure. The MSC are calculated from the ODS by the central difference approximation at grid point i

$$DI_{i\ MSCS}^{n} = \left| \left(\frac{\partial^2 w^n}{\partial x^2} \right)_i^2 - \left(\kappa_x^n \right)_i^2 \right| = \left| \left(\frac{\left(w_{i+1}^n - 2w_i^n + w_{i-1}^n \right)}{h^2} \right)^2 - \left(\kappa_x^n \right)_i^2 \right| \quad (7),$$

where w^n is a measured transverse displacement of the structure, κ_x^n is reconstruction of Fourier series approximation of MSC data in x direction, n is a ODS number, i is number of grid point in x direction and h is the distance between two successive measured points. The maximum value indicates the location of damage. Afterwards, this DI is summarized, standardized and DER value is calculated, according to equation (3), equation (4) and equation (5), respectively.

3. Specimens and experimental set-up

Two laminated composite beams considered in this study are cut out from a carbon/epoxy composite plate with The laminate lay-up $(0/90/+45/-45)_s$ and ply thickness of t = 0.3 mm, thus thickness of the plate is 2.4 mm. Experimentally determined material properties of the plate are as follows: E_x = 54.5 GPa, E_y = 31.04 GPa, G_{xy} = 7.09 GPa, G_{yz} = 6.5 GPa, ν_{xy} = 0.3, ρ = 1364.9 kg/m^3. Dimensions of the beams: width B = 35 mm and thickness H = 2.4 mm for both beams, Beam 1 – length L = 550 mm, Beam 2 – L = 350 mm.

Impact tests were performed on INSTRON Dynatup 9250 HV drop tower in figure 1 (left). By varying the drop height, different impact energies were obtained. The beams were fixed by using pneumatic clamps. For the Beam 1 the impact energy of 15 J was selected, location of impact is set at the distance L_i = 345 mm. For the Beam 2 – 10 J at the distance L_i = 175 mm.

The resonant frequencies of the beams are measured by a POLYTEC PSV-400-B scanning laser vibrometer (figure 1 (right)). PSV system requires the definition of outer edges and scanning grid, which consists of points where vibrations are measured. 56 and 30 equally spaced scanning points are equally distributed to cover area of *Beam 1* and *Beam 2*, respectively. Specimens were tested under clamped-clamped (two ends fixed) boundary conditions, where two vices were used to fix the ends of the beam (10 mm from both sides) with the clamped torque of 20 Nm. The beams are excited by a periodic input chirp signal, generated by the internal generator with a 4800 Hz bandwidth through a piezoelectric actuator discs. Afterwards, the beam is excited by a periodic sine wave signal with a frequency, equal to each resonant frequency to obtain the ODS by taking the Fast Fourier Transform of the response signal.

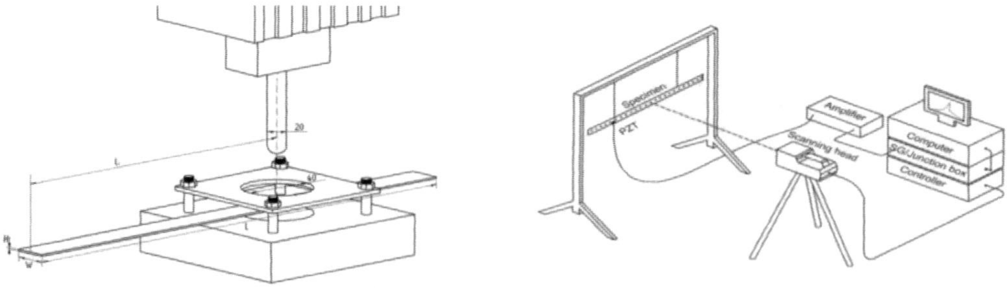

Figure 1. Left: low-velocity impact testing, right: experimental set - up.

4. Results and discussion

SDI values were calculated for wavelets functions, mentioned in section 2.1 for scales s = from 1 to 128. For each of these 128 SDI values one DER value from equation (5) was calculated. DER values were also calculated, using MSCS method. Fourier series approximations are depicted with F2-F8. Respective results are shown in table 1.

IOP Conf. Series: Materials Science and Engineering **111** (2016) 012005 doi:10.1088/1757-899X/111/1/012005

Table 1. DER values for approximation functions and wavelets with highest maximum DER.

CWT		MSCS		DER (%)				Scale	
Wavelets		Approx. functions		Beam 1		Beam 2		Beam 1	Beam 2
Beam 1	Beam 2	Beam1	Beam 2	CWT	MSCS	CWT	MSCS	3	4
cMor2	cMor2	F4	F3	90.2	88.8	78.8	86.7		
With threshold of 1.28				95.4	97.5	89.6	100		

It was necessary to compute DER versus scale plots in order to determine the value of scale parameter at which damage identification was the most successful. As one can see in figure 2, DER peak values correspond to regions of finer scales. After the first peak, the rest of the DER(s) pattern is unpredictable – it resembles a noisy signal. Nevertheless, these plots DER(s) are unique to each wavelet and specimen under test. Only results for wavelets with highest maximum DER values are shown. Maximum DER values for Beam 1 are better by about 11 % (2 %) than respective results for Beam 2 in CWT (MSCS). The best wavelet function turned out to be complex Morlet function of order 2 for both beams. As for MSCS method, different Fourier functions performed differently – Fourier function of order 4 (F4) for Beam 1 was the best, whereas F3 for was the best for Beam 2.

Damage identification results for both beams are shown in figure 3 for CWT and MSCS methods. As one can see, the location of damage, which is enclosed by two red dashed lines, is successfully identified by both methods. There are some smaller peaks present in these plots, which are later filtered out by application of SDI threshold value of 1.28 ($SDI_{1.28}$ in the plot).

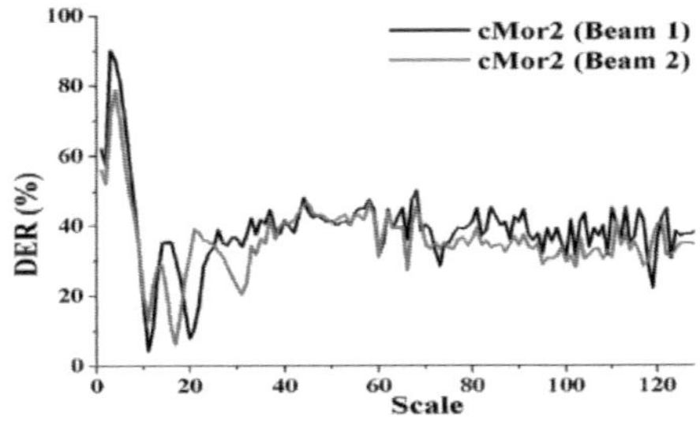

Figure 2. DER (s) plots for both beams.

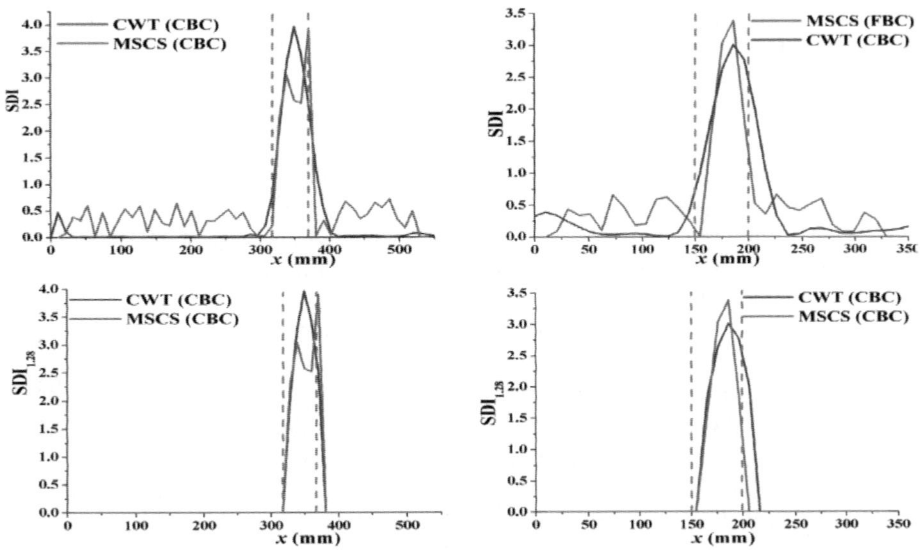

Figure 3. SDI (x) plots. Left – Beam 1, right – Beam 2. Up – no threshold, bottom – threshold of 1.28.

5. Conclusions

In the present paper, an experimental study on damaged location in two carbon/epoxy polymer composite beams, subjected to low-velocity impact damage, was conducted, exploiting two methods, namely, spatial CWT and MSCS. Success of damage detection was demonstrated using statistical hypothesis approach with a truncation of smaller magnitude peaks in SDI, thus obtaining damage localization with a confidence of 90 %. The reliability of damage identification was determined via DER value, which was calculated for 78 different wavelet functions, each of which at scales from 1 to 128 and for Fourier approximation functions of orders 2 to 8 for MSCS method.

While both methods – CWT and MSCS offer an advantage that there is no need for ODS information of a healthy structure and their DER values are comparable, a significant benefit of MSCS method over CWT is that damage identification algorithm is less cumbersome – there is no need to analyse results at different parameter values (scales for wavelets) and there is a considerably smaller variation of available functions for MSCS method as opposed to large amount of wavelet functions. Also, note that only best results were shown, meaning that there is a considerable variation in DER values for wavelets. However, in real life situations, when location of damage in not known a priori, one must still perform SDI calculations for different functions and compare damage detection results, relying on the results that are similar among majority of functions, while dismissing the ones, which have located damage in a completely different position.

Acknowledgement

The research leading to these results has received the funding from Latvia State Research Programme under grant agreement "Innovative Materials and Smart Technologies for Environmental Safety, IMATEH".

References

[1] Rucevskis S, Wesolowski M and Chate A 2009 Damage detection in laminated composite beam by using vibration data *J. Vibroeng.* **11** 363–73.
[2] Jerri A J 2011 *Introduction to Wavelets, first edition* (Potsdam, New York: Sampling Publishing) pp 417-41.
[3] Wu N and Wang Q 2011 Experimental studies on damage detection of beam structures with wavelet transform *Int. J. Eng. Sci.* **49** 253-61.
[4] Kim H and Melhem H 2004 Damage detection of structures by wavelet analysis *Eng. Struct.* **26** 347–62.
[5] Katunin A and Holewik F 2013 Crack identification in composite elements with non-linear geometry using spatial wavelet transform *Arch. Civ. Mech. Eng.* **13** 287-96.
[6] Katunin A 2014 Vibration-based spatial damage identification in honeycomb-core sandwich composite structures using wavelet analysis *Compos. Struct.* **118** 385–91.
[7] Katunin A 2015 Stone impact damage identification in composite plates using modal data and quincunx wavelet analysis *Arch. Civ. Mech. Eng.***15** 251–61.
[8] Gallego A, Moreno-Garcı´a P and Casanova Cesar F 2013 Modal analysis of delaminated composite plates using the finite element method and damage detection via combined Ritz/2D-wavelet analysis *J. Sound Vib.* **332** 2971–83.
[9] Solis M, Algaba M and Galvin P 2013 Continuous wavelet analysis of mode shapes differences for damage detection *Mech. Syst. Signal Pr.* **40** 645 - 66.
[10] Gentile A and Messina A 2003 On the continuous wavelet transforms applied to discrete vibration data for detecting open cracks in damaged beams *Int. J Solids Struct.* **40** 295–15.
[11] Bayissaa W L, Haritosa N and Thelandersson S 2008 Vibration-based structural damage identification using wavelet transform *Mech. Syst. Signal Pr.* **22** 1194–15.
[12] Matlab R2014a: http://se.mathworks.com/help/curvefit/fourier.html.

IOP Conf. Series: Materials Science and Engineering **111** (2016) 012006 doi:10.1088/1757-899X/111/1/012006

Search for composition of nanoparticles containing poly (vinyl alcohol)/poly (vinyl acetate) blend composites having the highest value of the modulus of elasticity by the response surface methodology.

A Kovalovs[1], N Jelinska[2], M Kalnins[2], A Chate[1]
[1]Riga Technical University, Faculty of Civil Engineering, Institute of Materials and Structures, Kipsalas Street, 6B/6A, Riga, LV-1048, Latvia
[2]Riga Technical University, Faculty of Material Science and Applied Chemistry, Institute of Polymer Materials, P. Valdena Street 3, Riga, LV-1048, Latvia

E-mail: andrejs.kovalovs@rtu.lv

Abstract. The aim of this study was to investigate the dependence of elastic modulus on component's proportion of nanoparticles containing PVA/PVAc blend composites by means of response surface methodology. Full factorial design method was used to perform a series of experiment. The statistical analysis of experimental results showed that adding of nanoparticles to the PVA/PVAc compositions has substantial effect on the tensile properties. The higher values of the tensile elastic modulus show systems with almost equal content of polymers in the blend. Regardless of PVA/PVAc content relationships, the higher modulus values present systems that contain both nanoparticles. Composition, which provide greatest value of elastic modulus was determined and experimentally verified.

1. Introduction
Poly(vinyl alcohol) (PVA) is the it is the world's most produced synthetic water-soluble polymer. It is produced commercially by the hydrolysis of poly vinyl acetate (PVAc). To improve mechanical and other properties PVA is blended with different synthetic polymers. Composites prepared by blending of comparatively rigid poly vinyl alcohol (PVA) with rather compliant poly vinyl acetate (PVAc) show attractive strength-deformation characteristics which change significantly with PVA/PVAc proportions [1]. For example, the elastic modulus of blends changes up to seven times, from 0,3 GPa to 2,2 GPa. It seemed interesting to strengthen PVA/PVAc compositions by addition of some particles with reinforcing effect: placoid particles of montmorillonite (MMT) [2] and particles of microcrystalline cellulose (MC) [3]. Efficiency of presence of small amounts (less than 5%) of such nanoparticles on strength-deformation characteristics of number of thermoplastic polymers is proven frequently [4-6].

Application of response surface methodology (RSM) allows finding out the relationship between components proportions and tensile characteristics of composite significantly reducing the number of experiments [7]. The RSM is widely used to optimize process parameters, especially in determining optimum conditions for chemical investigations. The helpfulness of RSM in the development and optimization of a variety of composites has been highlighted by several authors [8].

The aim of this study was to investigate the dependence of elastic modulus on component's proportion of PVA/PVAc blend composites containing nanoparticles of montmorillonite clay (MMT), microcrystalline cellulose (MC) by means of RSM and to find the composite composition with greatest value of elastic modulus.

2. Materials and method

2.1. *Optimisation procedure*

The general flowchart of the optimisation procedure that should be followed in the present research is shown in Figure 1. The optimisation process consists of six stages. In the first stage (Experimental Design), a plan of experiments is produced according to the number of design variables and number of experiments. In the second stage (Test), the specimens are created in order to model the response of a structure and testing is performed in the reference points of the experimental design. In the third stage (Approximation), the numerical data obtained by the testing of specimens in the sample points is used in order to build the approximating functions using the response surface method. These simple mathematical models obtained from the data of experiments are used as the objective functions in the optimal design problem. The non-linear optimisation problem is executed by the random search method using the obtained response surfaces in the next stage (Optimisation). The optimal result of non-linear optimisation is checked using the testing in the fifth stage (Result validation). The optimal design variables are used for preparation of the specimen in order to compare difference between the optimisation result and the experimental solution. If the difference between the optimal result and the experimental solution is higher than 5% it is needed to improve the correlations of the approximating functions or change the design space for some parameters. The procedures are executed before obtaining the final optimal solution.

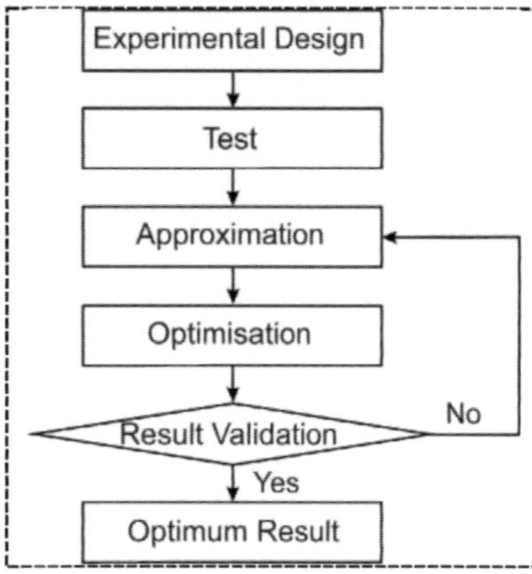

Figure 1. Flowchart of the optimisation procedure.

2.2. *Materials*

PVA (from *Merck Schuchardt*) (molecular weight of 133,000 Da, content of acetate groups 0.98%), as well as a PVAc water emulsion (Mowilith DHS S1, from *Celanese Emulsions GmgH*) with solid content 50% were used as initial materials. Diethylene glycol (*Sigma-Aldrich*) was used as plasticizer for both polymer components.

Montmorillonite clay (from *Dellite, Laviosa Chimica Mineraria*) with dimensions of particle size after dispersion in water 1x500 nm and microcrystalline cellulose as 9% water gel were used. A 10% water solution of PVA was mixed with a 10% PVAc water emulsion in different proportions providing the specified volume fractions φ of the components in the blend. MMT was added to polymers in a form of 4% water dispersion. Dispersion was prepared by mixing for 8 hours (magnetic stirrer) followed by ultrasonic technique for 10 minutes.

The films were obtained by casting: the mixtures were poured into *Petri* dishes 140 mm in diameter (the initial thickness of the layer was 2 mm) and dried at a temperature of 17°C up to a constant content of water about 3%. The thickness of the films obtained was 150 ± 10 μm.

2.3. *Mechanical test*

The strength values were obtained according to EN ISO 527 standard of Determination of tensile properties and were measured by Zwick/Roell universal machine at constant crosshead speed 20 mm/min at room temperature.

The data obtained were processed by using the testXpert V11.0 program. The stress–strain relationship $\sigma(\varepsilon)$ was determined up to the failure of specimens. Each set of experimental design was replicated five times and the average values of initial elasticity modulus $E = \lim(d\sigma/d\varepsilon)|_{\varepsilon\to 0}$ were determined. The average error of tensile elastic modulus values did not exceed ± 10 %.

2.4. Experimental design

The experiments were designed according to Full Factorial design (FFD). FFD of experiment is the most popular designs owing to their simplicity and relatively low cost. It is very useful for preliminary studies or in initial optimization steps. A FFD contains all the combination of the different levels of all the design variables.

The plan of experiments is formulated for 3 design variables, namely, nanoparticles of montmorillonite clay (X_1), nanoparticles of microcrystalline cellulose (X_2), volume fraction of PVA in PVA/PVAc blend (X_3) and 27 experiment points. The experimental plan with coded and uncoded levels of design factors and experimental results are given in Table 1.

Table 1. Full factorial design plan and experimental results.

Experiment No.	Coded variables			Actual variables			Response (Elastic modulus, MPa)
	X_1	X_2	X_3	x_1	x_2	x_3	Y
1	1	1	1	5	5	90	434
2	1	1	0	5	5	50	453
3	1	1	-1	5	5	10	407
4	1	0	1	5	2.5	90	401
5	1	0	0	5	2.5	50	418
6	1	0	-1	5	2.5	10	366
7	1	-1	1	5	0	90	362
8	1	-1	0	5	0	50	370
9	1	-1	-1	5	0	10	320
10	0	1	1	2.5	5	90	406
11	0	1	0	2.5	5	50	421
12	0	1	-1	2.5	5	10	369
13	0	0	1	2.5	2.5	90	412
14	0	0	0	2.5	2.5	50	441
15	0	0	-1	2.5	2.5	10	376
16	0	-1	1	2.5	0	90	356
17	0	-1	0	2.5	0	50	368
18	0	-1	-1	2.5	0	10	316
19	-1	1	1	0	5	90	348
20	-1	1	0	0	5	50	366
21	-1	1	-1	0	5	10	310
22	-1	0	1	0	2.5	90	337
23	-1	0	0	0	2.5	50	363
24	-1	0	-1	0	2.5	10	307
25	-1	-1	1	0	0	90	313
26	-1	-1	0	0	0	50	344
27	-1	-1	-1	0	0	10	288

Subsequently in the points of plan of experiment the results of testing were obtained. In the next stage the numerical data obtained by the testing in the points of plan of experiments was used in order to build the approximating functions.

Eglais [9] proposed the algorithm of selection of the terms for non–polynomial regression function. In his work, the "correlation" approach for the optimal choice of the number of terms was also proposed. This approximation method proposed by Eglais is employed for the solution of the optimisation problem considered here. In the present approach the form of the regression equation is unknown in advance. There are two requirements for the regression equation: accuracy and reliability. Accuracy is characterised as a minimum of standard deviation of the table data from the values given by the regression equation. It is possible to obtain a complete agreement between the table data and values given by the regression equation, by increase of the number of terms in the regression equation. However, it is necessary to note that prediction in the intervals between the table points may not be accurate.

For an improvement of prediction, it is necessary to decrease the distance between the points of the experiments by increasing the number of experiments or by decreasing the domain of factors. Reliability of the regression equation can be characterised by the affirmation that standard deviations for the table points and for any other points are approximately the same. Obviously, the reliability is greater for a smaller number of terms of the regression equation.

The regression equation can be written in the following form:

$$y = \sum_{i=1}^{p} A_i f_i(x_j) \tag{1}$$

where Ai are the coefficients of the equation of regression, fi(xi) are the functions from the bank of simple functions φ_1, φ_2,...φ_K which are assumed as

$$\varphi_k(x_j) = \prod_{i=1}^{n} x_j^{\alpha_{ki}} \tag{2}$$

where n is the number of the object parameters and α_{ki} is a positive or negative integer including zero. Synthesis of the equation from the bank of simple functions is carried out in two stages: selection of perspective functions from the bank and then step-by-step elimination of the selected functions.

In the first stage, all variants are tested with the least square method and the function, which leads to the minimum of the sum of deviations, is chosen for each variant. In the second stage, the elimination is carried out using the standard deviation:

$$\sigma_0 = \sqrt{\frac{S}{k-p+1}}\ ,$$
$$\sigma = \sqrt{\frac{1}{k-1}\sum_{i=1}^{k}\left(y_i - \frac{1}{k}\sum_{j=1}^{k} y_j\right)^2} \tag{3}$$

or correlation coefficient:

$$c = \left(1 - \frac{\sigma}{\sigma_0}\right) * 100\% \tag{4}$$

where k is the number of experimental points, p is the number of the selected perspective functions and S is the minimum sum of deviations. It is more convenient to characterise the accuracy of the regression equation by the correlation coefficient.

3. Results and discussion

Relationship between the design variables x and the corresponding behavior functions Y is given as (correlation $c = 93.4\%$):

$$Y = 262.7 + 27.74x_1 + 20.94x_2 + 2.499x_3 + 2.093x_1x_2 - 4.124x_1^2 - 3.11x_2^2 - 0.02x_3^2 \tag{5}$$

After selection of equation of regression the parametric studies were carried out additionally to scrutinize the influence of design parameters on behaviour function. This was done by displaying 3D graphs of approximating functions. The view of the response surfaces is shown in Figure 2. This figure shows the estimated response function for the different amount of nanoparticles with different combinations of volume fraction of PVA.

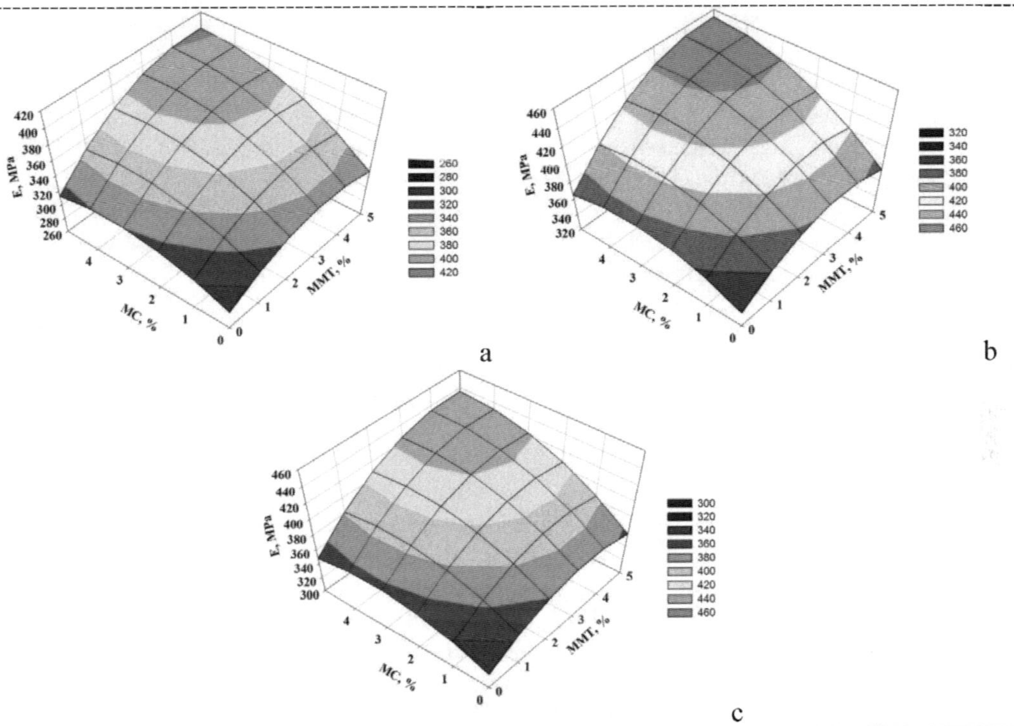

Figure 2. Dependence of elastic modulus E on of nanoparticles content (volume fraction of PVA: a -10%, b – 50% and c – 90%).

It can be seen that the value of elastic modulus increases proportionally to an increase of nanoparticles. When the amount of nanoparticles begins to grow, the elastic modulus increase significantly up to a certain point and then increases slightly. In according to individual main effects of design variables, the most effective factor on elastic modulus is nanoparticles of montmorillonite clay. Figure 2b shows that maximum value of elastic modulus is achieved when amount of PVA and PVAc is equal.

4. Optimisation

The search of the optimal composition of PVA/PVAc blend films on the basis of the mathematical model was the final stage of this research. Non-linear optimisation problem was solved by the random search method using the response surface obtained in the corresponding program EdaOpt [10]. The optimisation result obtained for PVA/PVAc blend film is outlined in Table 2.The maximum elastic modulus reaches 454 MPa, when content of montmorillonite clay is 4.5%, microcrystalline cellulose is 5.0 % and volume fraction of PVA in the blend PVA/PVAc is 60%. The optimal result obtained with the response surface method was compared with the experimental result in the point of optimal solution. It is seen that differences between the optimal and experimental results using nanoparticles to the PVA/PVAc compositions obtained by the optimisation are very small. Mostly residuals do not exceed 2.5%, which show good correlation of the approximating functions.

Table 2. Optimal solution for a maximal elastic modulus.

Title	Design variables			Objective function
	x_1	x_2	x_3	Y
Optimisation	4.5	5.0	60	454
Experimental verification	4.5	5.0	60	443

5. Conclusion

The effect of the nanoparticles of montmorillonite clay, nanoparticles of microcrystalline cellulose and PVA/PVAc ratio on the elastic modulus of PVA/PVAc blends was investigated using response surface method. Data obtained and 3D graphs show that adding nanoparticles to the PVA/PVAc compositions has a substantial effect on the tensile properties. Irrespective of type of nanoparticles, the higher values of tensile elastic modulus show systems with equal content of polymers in the blend. Regardless of PVA/PVAc content relationships, the higher modulus values present systems that contain both kinds' nanoparticles. Composition, which provide greatest value of elastic modulus was determined and experimentally verified.

Acknowledgments

The present work was supported by the European Social Fund within the project 2013/0017/1DP/1.1.1.2.0/13/APIA/VIAA/06

References

[1] Jelinska N, Kalnins M 2011 Strenght and deformation characteristics of polymer blend films obtained from water systems *Mechanics of Composite Material* **47**(5) 236-250

[2] Pavlidou S, Papaspyrides CD 2008 A review on polymer–layered silicate nanocomposites *Progress in Polymer Science* **33**(12), 1119–1198

[3] Laka M, Chernyavskaya S 2007 Obtaining and properties of microcrystalline cellulose from hardwood pulp *RTU Material Sciences and Applied Chemistry* **14** 7-14

[4] Sinha Ray S, Okamoto M 2003 Polymer/layered silicate nanocomposites: a review from preparation to processing *Progress in Polymer Science* **28**(11) 1539–1641

[5] Ma J, Xu J, Ren J, Yu Z, Mai Y 2003 A new approach to polymer/montmorillonite nanocomposites *Polymer* **44**(16) 4619–4624

[6] Tjong SC 2006 Structural and mechanical properties of polymer nanocomposites *Materials Science and Engineerin* **53**(3-4) 73–197

[7] Montgomery DC 2012 Design and Analysis of Experiments (NY: Wiley)

[8] McCarron PA, Woolfson AD, Keating SM 2000 Response surface methodology as a predictive tool for determining the effects of preparation conditions on the physicochemical properties of poly(isobutylcyanoacrylate) nanoparticles **193** 37–47

[9] Eglais V 1981 Approximation of data by multi–dimensional equation of regression *Problems of dynamics and strength* 39 120–125

[10] Auzins J, Janushevskis A, Janushevskis J, Skukis E 2014 Software EdaOpt for experimental design, analysis and multiobjective robust optimization *Proc. OPT-i Int. Conf. Engineering and Applied Science Optimization (Kos Island, Greece, 4–6 June 2014)* (Athens: National Technical University) pp 101-123

Solvent pre-treatment of lignin for obtaining of low viscosity lignopolyols by oxypropylation for synthesis of polyurethanes

A Jablonskis, A Arshanitsa and G Telysheva

Latvian State Institute of Wood Chemistry, Riga, Latvia

E-mail: antonio252@inbox.lv

Abstract. Wheat straw lignin was pre-treated by organic solvents of different polarity: ethylacetate, ethylmethylketone and ethylmethylketone/water mixture, using sequential extraction technique. The total yield of fractions was ~ 40% (w/w). In comparison with non-treated lignin the solvent soluble fractions have lower molecular weights, lower polydispersity and average functionality indices. The low viscosity lignopolyols with lignin substituent content up to 40%, suitable for rigid polyurethane foams obtaining were produced by the oxypropylation of isolated fractions. The non-soluble in organic solvents lignin residue represent a purified natural aromatic amorphous cross-linked polymer. Therefore besides oxypropylation its application as filler of thermostable polymers with enhanced flammable resistance could be beneficial biorefinery approach.

1. Introduction

Lignin comprising 15 – 25% of lignocellulosic feedstocks such as agricultural materials or forest resources, offer a number of attractive features as starting material for production of polymers and chemicals of aromatic nature.

The difficulties with lignin conversion into value added products is connected with high intrinsic heterogeneity of lignin in situ and its strengthening in the process of their isolation from plant tissue. Structural and functional heterogeneity of lignin is conditioned by the diversity of phenyl – propan moieties forming of lignin structure and combination of them in different ratio depending on plant origin, type of links between moieties, cross linking and severity of processes used for lignin isolation. Different approaches were used for overcoming negative impact of lignin heterogeneity incl. structural moieties (by introduction of novel additional functional groups and extension of chain) with respect to selected task and fractionation by different methods: ultrafiltration, selective precipitation, permeation chromatography.

The object of this study was combination of above mentioned approaches for obtaining of liquid polyols with enhanced content of lignin prospective for PU synthesis from pure soluble technical wheat straw lignin. The sequential extraction of lignin with solvent of increasing hydrogen – bonding capacity was used for pretreatment and obtaining of lignin fractions of reduced heterogeneity, meanwhile for following modification of fractions was realized using propylene oxide, that provide introduction of high active in reaction with isocyanate secondary OH- groups in exchange of phenolic one [1].

2. Experimental

2.1. Materials

BIOLIGNIN™ was extracted at pilot scale from wheat straw using a mixture of acetic acid/formic acid/ water at the CIMV pilot plant (Pomacle, France). Purity of the lignin was about 95% [2]. Initial BIOLIGNIN™ was washed with water to set free lignin from organic acids, then dried at 50°C and grinded. Fractions, isolated by sequential fractionation [3] of initial BIOLIGNIN™ (F0) using ethylacetate (F1), ethylmethylketone (F2) and ethylmethylketone/ water (90/10, v/v) mixture (F3) and residue (F4) were dried in oven at 70°C and after that fractions was dried in vacuum at 50°C.

2.2. Oxypropylation reaction conditions

The initial lignin (L) and isolated fractions were oxypropylated in 1L Parr high pressure reactor. Lignopolyols, differing by the lignin content in the initial reaction mixture from 10 to 40% (based on the dry mass of lignin), were synthesized. In all cases propylene oxide (PO), air dried lignin (5% moisture) and KOH (5% based on the dry mass of lignin) were loaded into the reactor, sealed and heated under stirring to 160 – 176 °C, when exothermic process started then heating is switched off. The reactor pressure increased and then dropped to a value close to atmospheric pressure. Showing that the reaction is over, the KOH was neutralized by acetic acid. The ensuing products were recovered of the unreacted PO by vacuum removal [3].

2.3. Characterization of lignopolyols obtained

The viscosities of lignopolyols was determined by means of rotation viscosimetry [HAAKE Viscotester 6L/R plus] at 25 ° C. The hydroxyl index (I_{OH}) were determined via acetylation of samples with Ac_2O followed by potentiometric titration of free acetic acid in excess with 0.1 N NaOH in water. Potassium acetate (KAc) content in lignopolyols (ethanol solution) was determinated by potentiometric titration (Radiometer analytical TIM 840 equipped with glass electrode) with 0.1 N HCl. The moisture of lignopolyols obtained was determinated using Carl Fisher titration Culonometric KF titrator 275 KF

2.4. Characteristics of Lignin-PO copolymers [3]

Liquid – liquid extraction was used for lignin copolymer isolation from reaction mixture. Lignopolyol was dissolved in methanol and then hexane was added. The mixture was stirred 3h and then hexane phase was separated. Extraction with hexane was repeated one more time. Hexane was distilled and high molecular polypropylene glycols (PPG) content was determined gravimetrically. Methanol solution was concentrated using rotation evaporator. Concentrated methanol solution was precipitated in cold acidic water. Lignin - PO copolymer was filtered, washed and dried.

Lignin - PO copolymer glass transition (T_g) temperature was determined by DSC analysis using with Mettler Toledo Star DSC823 instrument. Scans were recorded from -70°C to 200°C under N_2 atmosphere with a heating rate of 10 °C/min in an alumina crucible. For each experiment, 8mg dried specimens were used.

Determination of methoxyl and oxypropyl group content, is based on the Baker's method [3] where final step – detection of reaction products containing methyl iodide and isoiodpropane, was change on chromatograph. The samples was mixed with hydroiodic acid in a vial with a pressure-resistant seal and heated in oil bath. The vial was cooled in ice–water and chloroform containing 1-iodpropane as an internal standard (IS) was added immediately with a syringe through the sealed cap. The reaction products and the IS were extracted with chloroform, dried over anhydrous sodium sulphate, and subjected to GC analysis.

The ratio of the peak areas of methyl iodide, isopropyl iodide to that of IS was calculated by the Agilent 6850 software. The system was calibrated with vanillin.

AU (arbitrary units): a ratio of peak area of isopropyl iodide to that of the internal standard on chromatograms.

3. Results and discussion

3.1 The characteristics of lignin fractions

The BIOLIGNIN^{TM} is not fully soluble in most common organic solvents (tetrahydrofuran, dioxane, acetone, ethanol, methanol, ethyl acetate and other). Sequential extraction used ethylacetate, ethylmethylketone and ethylmethylketone/water mixture allow to isolate 36% of lignin as organo-soluble fraction (the content of lignin residue (F4) after solvent pre-treatment was 64,1%) characterized by 3 times lower heterogeneity (Mw/Mn) and molecular mass and decreased of OH groups per macromolecule unit (average functionality). In the result of fractionation was obtained 3 soluble lignin fractions with reduced heterogeneity and functionality (Table 1).

Table 1.

Characteristics of BIOLIGNIN^{TM} and isolated organosoluble fractions

Sample	Yield, %	Mn, g/mol	Mw, g/mol	Mw/Mn	M_{EQ}, g/mol	Average functionality
BIOLIGNIN^{TM} (F0)	100	900	17800	19,8	122	7,3
Ethylacetate fraction (F1)	11,2	270	1500	5,4	236	1,2
Ethylmethylketone fraction (F2)	9,2	380	2500	6,6	148	2,7
Ethylmethylketone/ water fraction (F3)	15,5	670	4500	6,7	135	4,9

- The data of molecular weight distribution were kindly presented by RG Gosselink (Wageninging, The Netherlands)
  - Mn – number average molecular weight;
  - Mw – weight average molecular weight;
  - Mw/Mn - polydispersity index (PDI);
  - M_{EQ} – amount of material containing 1mol of functional groups (OH groups).

It was suggested that ethylacetate (F1) and ethylmethylketone (F2) fractions with lower functionality have to give lower cross-linked polyurethanes (PU) in comparison with others samples.

3.2. Lignins oxypropylation and characteristics of lignopolyols obtained

The oxypropylation of BIOLIGNIN^{TM} and separated fractions made at different lignin/PO ratio (L/(L+PO)) (Table 2). As the result of oxypropylation reaction lignins the phenolic OH groups were transformed into secondary aliphatic OH groups high reactive towards isocyanate. Lignopolyols obtained from organic soluble fractions did not contain non-liquefied solid fractions that is significant advantage of this fractions as starting raw material for lignopolyols synthesis in comparison with unfractionated lignin (F0) or residue (F4).

Increase of lignin content in reaction mixture entail increasing of hydroxyl value (I_{OH}) of polyols obtained providing higher activity in PU formation. Together with that increasing of potassium acetate content in obtained lignopolyols with increasing of lignin content in them was observed. The presence of catalyst (KAc) active in polyurethanes and isocyanurate formation can be considered as an advantage in preparation of high active polyurethane systems usable for processing them into rigid polyurethane (PUR) foams (Table 2).

The water content in lignopolyols is varied in the range of 0.02-0.4% that meet the requirements to polyetherpolyols suitable for PUR foams production [5].

Table 2.

Characteristics of lignopolyols obtained

Sample	L/(L+PO), (w/w), %	Nonsoluble fraction,%	I_{OH}, mgKOH/g	Moisture $\times 10^{-2}$ %	KAc content, %	Content %	
						Copolymer	PPG*
F0	10	1-5	245±5	6.2±0.5	1,10±0,01	21,1±1,7	25,9±1,9
	20		358,5±2,2	6.7±0.5	1,75±0,03	32,4±0,7	16,5±1,9
	30		411±5	8.00±0.20	2,44±0,07	39,10±0,10	11,8±0,6
F1	10	0,0	168±6	2.90±0.10	1,01±0,01	42.9±2.4	33,2±1,0
	20		228±4	6.8±0.3	1,56±0,01	31,9±1,7	24,70±0,20
	30		315±7	3.0±0.6	2,17±0,10	35±4	26,4±1,8
	40		352,3±1,7	4.70±0.10	2,60±0,22	38,4±2,1	23,3±1,1
F2	10	0,0	172,2±0,6	1.90±0.10	0,83±0,01	20,1±1,2	19±3
	20		254,7±2,2	2.40±0.10	1,48±0,01	50±8	20,9±1,4
	30		330,4±2,2	5.8±0.20	2,20±0,10	40.0±1.0	15±3
	40		394±4	3.30±0.10	2,86±0,02	47,3±0,20	11,1±0,7
F3	10	0,0	196±17	8.90±0.10	0,89±0,01	19±3	30±3
	20		320±12	8.00±0.10	1,64±0,01	34,7±1,5	23,7±2,6
	30		411,2±1,1	9.30±0.10	2,5±0,5	39,0±1.0	16,5±1,8
	40		463,9±1,7	5.2±0.3	3,05±0,05	41±3	9,4±0,6
F4	10	0,5-3	234±3	7.5±0.4	0,85±0,02	15,8±1,8	22,1±0,8
	20		357±5	11.70±0.10	1,69±0,02	37±6	15,7±0,9
	30		415±6	40.4±3.3	2,50±0,10	41±6	12,30±0,10

* - High molecular hexane soluble polypropylene glycols

The lignopolyols synthesized on the basis of organic solvents soluble lignin fractions characterized with lower viscosity in comparison with viscosity of unfractionated lignin based polyols at the same lignin content in its (Figure 1).

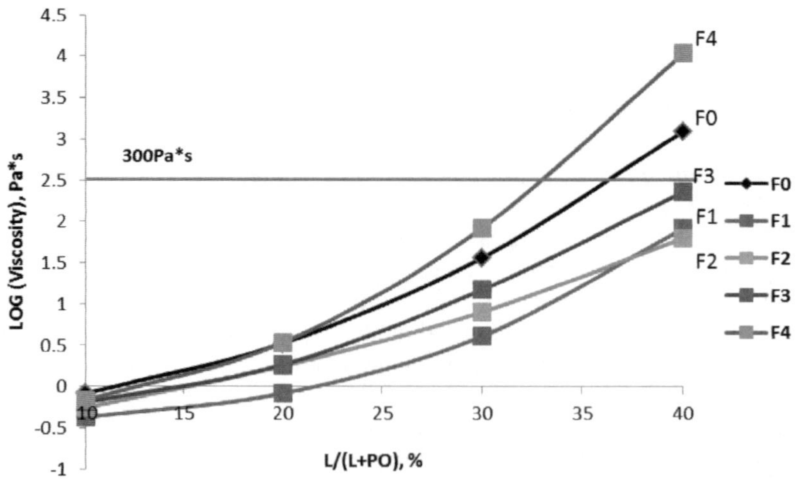

Figure 1. Viscosity of lignopolyols synthesized on the basis of BIOLIGNIN™ and its fractions

It is recognized that viscosity is one of the main characteristics of polyols determining the suitability for rigid polyurethane foam production in industry. Lignin component content in reaction mixture can be increased up 40% not exceeding 300Pa*s recognized as a limit viscosity for polyols in PUR foam industry [6, 7]. The results obtained allow to suggest the possibility to increase the lignin constituent content in lignopolyols up to 50%.

3.3. Lignin-propylene oxide copolymer characteristics

It was established that the lignopolyols obtained consist from lignin - PO copolymer, high molecular PPG soluble in hexane and low molecular PPG soluble in water. With increasing of L/(L+PO) values the lignopolyols obtained were steadily enriched in copolymer fractions (Table 2).

The mass balance of process of lignopolyols extraction indicated also the presence in lignopolyols of propylenglycol and its low molecular derivatives soluble in water, but insoluble in hexane. To evaluate the extent of lignin modification in results of oxypropylation the separated lignin - PO copolymers properties were studied in more details.

GC analysis of products obtained at copolymers treatments with HI has shown that the aliphatic OH groups in copolymers are represented mainly by secondary alcohol groups. The methoxyl and isopropyl groups of copolymers were determined in one attempt. It was found that copolymer methoxyl group amount increases, with increasing of L/(L+PO) ratio.

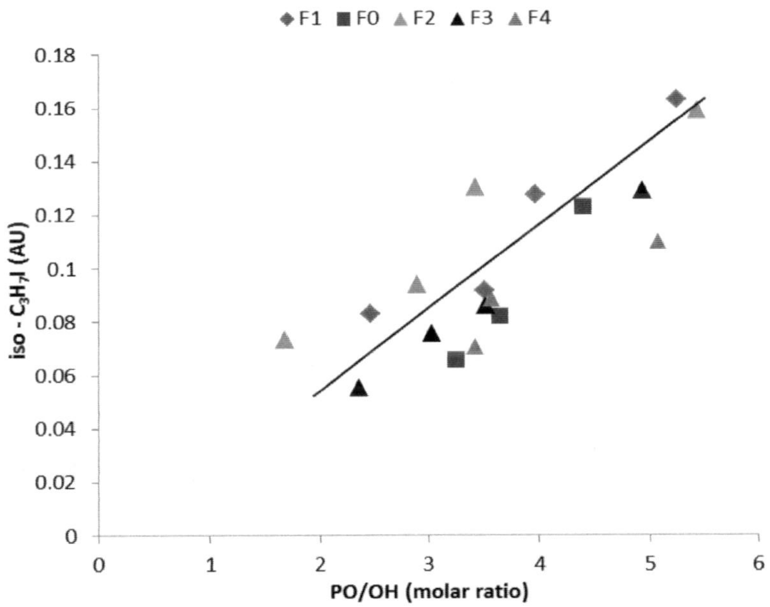

Figure 2. The isopropyl iodide content in products of copolymer treatment by HI vs PO units attached to one OH group of lignin

The growth of content of isopropyl groups in copolymers is accompanied by relative decreasing of -OCH_3 groups due to reduction of phenylpropane unit concentration in copolymers as the result of PO grafting on to OH functionality of lignins.

The calculation made on the basis of OCH_3 groups content in copolymers have shown , that per one OH group of lignin 2 till 5 PO molecules are grafted (Figure 2). The degree of molar substitution decreases in the following row F4>F2>F3>F1>F0.

The decreasing of lignin-PO copolymers Tg with increasing of isopropyl chain length of copolymer was established by DSC method, confirming the lignin modification

4. Conclusions

In contradiction to direct BIOLIGNIN™ oxypropylation, the fractionation and following oxypropylation of lignin fractions is the attractive approach to obtain lignopolyols with increased content up to 40% of renewables in composition, but with reduced viscosity, corresponding to industrial requirements ≤300 Pa*s. The lignopolyols obtained from fractions, isolated with organic solvent pre-treatment, do not contain non-liquefied solid fractions.

The lignopolyols obtained is consisting of lignin – PO copolymers, high and low molecular polypropylene glycols. PO grafting onto lignin OH groups was confirmed by independent physical-chemical methods: GC, FT-IR and DSC. The characterization of rigid polyurethane foams obtained from above mentioned lignopolyols is under development. The non-soluble in organic solvents lignin residue could be considered a purified natural aromatic amorphous cross-linked polymer. Therefore besides oxypropylation its application as filler of thermostable polymers with enhanced flammable resistance could be beneficial biorefinery approach. As an additional direction the carbonized material production using non-soluble lignin residue as a raw material could be proposed.

Acknowledgement

The financial support from 7th EU FP Collaborative project BIOCORE Contract 241566 is gratefully acknowledged.

Latvian budget State Research Program ResProd 2014-2017 is gratefully acknowledged.

References

[1] Bozell J J in book ''Plants and Bioenergy'' Advances in Plant Biology 4, edit. McCann M, Buckeridge M, Carpita N. *Springer Science+Business Media Ney York*, 2014, 277- 289.

[2] Delmas G H, Benjelloun-Mlayah B, Le Bigot Y, Delmas M 2011 *J. Appl. Polym. Sci.* **121**, 491-501.

[3] Arshanitsa A, Vevere L , Telysheva G, Dizhbite T, Gosselink J A R, Bikovens O and Jablonskis A 2015 *Holzforschung* **69(6)**, 785-793.

[4] Baker S M 1996 *Holzforschung* **50**, 573-574.

[5] Ionescu M 2005 *Rapra Technology Ltd.* (Shawbury-Shrewsbury-Shropshire,UK).

[6] Nadji H, Bruzzese C, Belgacem M.N, Gandini A, Benaboura A 2005 *Macromol. Mater. Eng.* **290**, 1009-1016.

[7] Cateto C A, Barreiro M F, Rodrigues A E, Belgacem M N 2009 *Ind. Eng. Chem. Res.* **48**, 2583-2589

Influence of monomer on structure, processing and application characteristics of UV curable urethane acrylate composite coatings

Z Grigale-Sorocina[1], M Kalnins[2] and K A Gross[2]

[1]Environment, Bioenergetics and Biotechnology Competence Centre, Aizkraukles street 21, 339, Riga, LV-1006, Latvia
[2]Riga Technical University, Faculty of Material Science and Applied Chemistry, Institute of Polymer Materials, P. Valdena Street 3, Riga, LV-1048, Latvia

E-mail: zane.grigale@gmail.com

Abstract. Increased interest in the esthetical natural nail coatings have encouraged more in-depth studies particularly of UV curable coatings: their formation, processing, structure, characteristics and removing. Typical requirement for nail coatings is good adhesion, but preferably for the short time of functioning (usual 2-4 weeks). This study investigated the impact of four different monomers (tertiobutyl cyclohexyl acrylate (TBCHA), ethylene glycol dimethacrylate (EGDMA), tetrahydrofurfuryl acrylate (THFA), hydroxypropyl methacrylate (HPMA)) to viscosity of uncured mixture and degree of conversion, mechanical properties, surface gloss, micro hardness and adhesion loss for cured films. Specific coating application requires comparatively high coating flexibility and stability of deformation characteristics. This can be achieved with composition containing 30% of monomer TBCHA, what shows ultimate elongation ε_B = 0.23 – 0.24, modulus of elasticity E = 670-710 MPa and comparatively constant properties in 72 hours (ΔE = 1.3%, $\Delta\varepsilon_B$=6.0%). A composition with 40% of TBCHA shows the fastest coating destruction achieving adhesion loss within 3 min.

1. Introduction

During last decades, a great deal of attention has been given to UV curing applications as an alternative to traditional solvent-borne coatings [1, 2, 3] due to their environmental soundness, lower energy consumption, rapid curing at room temperature and curing on heat sensitive substrates [4, 5, 6, 7]. In the last decade new market space for coating applications with short-term requirements has appeared. The market space includes coatings on natural nail for aesthetic and medical reasons [8, 9]. Typical coating requirement is good adhesion, but for these novel applications adhesion is desired only for the time of functionality (usual 2-4 weeks).

The formulation of appropriate coatings is critical to achieve required characteristics: firstly, for uncured gel - pigment stability of the suspension and reduced viscosity (3000-4500mPas) – secondly, for cured coating - adhesion loss from substrate (after 5-10 min soak off in the solvent permitted in cosmetics), increased elasticity (20-50%), high surface gloss (70-90), surface hardness (>110MPa) and efficient polymerization (90-95%), what can be achieved with accurate ingredient selection.

Short-term coating applications haven't been widely reported yet, only some patent applications have been submitted [10, 11, 12]. Desired ingredients in radically cured formulations are polyurethane acrylates as precursors to produce cross-linked networks giving high-performance final materials. As

UV curable resins they prove excellent adhesion, flexibility, chemical and scratch resistance, high hardness and gloss, but often suffer from the high viscosities [1, 4, 7]. To overcome the deficiency low viscosity monofunctional monomers or low viscosity oligomers as reactive diluents can be used. Degree of conversion of the cross-linked system is one of the most important parameters which can strongly affect adhesion loss and required short-term characteristics. Degree of conversion can be controlled with light intensity, time of curing and photoinitiator amount [4, 6, 13]. For coatings on natural nails all these 3 aspects are constant as temperature on natural nails cannot exceed 70°C. Curing time is usually 30 s in the lamp with UV spectrum 395-405nm and photoinitiator amount 1-2%. Other possibility to affect the degree of conversion is optimization of standard system. Some studies have been done by varying with different photoiniciators [4, 14], urethane acrylate oligomers [6] and monomers [14]. Biopolymers [15, 16, 17] and their effect on the short-term characteristics of UV curing coating have been studied by us previously [18, 19]. There still is a need to optimize the standard systems, varying with monomer type, functionality and its content what may accelerate the adhesion loss in the solvent due to the incomplete curing and less dense network. We put forward the hypothesis that less assembled systems and monomers with lower functionality provide weaker adhesion to the substrate and ensure the possibility to eliminate the adhesion interaction between the coating and the substrate.
The objective of this study is to optimize the composition of the UV curing varnish system composed of urethane acrylate oligomers, photoinitiator and monomers of various types and functionalities.

2. Experimental

Materials were selected to ensure two different goals: long term properties - with good adhesion and short term properties –quick removal in solvent media. For uncured gel viscosity was determined. For cured coatings degree of conversion, mechanical characteristics, surface gloss, microhardness and adhesion loss in solvent media were evaluated.

2.1. Materials and methods

Main composite ingredients (oligomers, monomers, photoinitiator) were selected to confirm the requirements of EU Cosmetic Regulation. Tri-functional urethane acrylate oligomer Ebecryl 1258 (*Allnex)* was chosen for its film forming properties. Hexa-functional urethane acrylate oligomer Ebecryl 1291 (*Allnex)* was added as adhesion promoter. Four different monomers were evaluated to study their effect on the degree of conversion and adhesion of the cured film: Tertiobutyl cyclohexyl acrylate (TBCHA), Ethylene glycol dimethacrylate (EGDMA), Tetrahydrofurfuryl acrylate (THFA), Hydroxypropyl methacrylate (HPMA). For the initiation of radical photopolymerization reactions Ethyl - 2,4,6 – Trimethylbenzoylphenylphosphinate (*Lucirin TPO_L*) (*BASF*) photoiniciator was applied.

2.2. *Composition preparation*

All components - reactive diluent/monomer (conc.10, 20, 25, 30, 35, 40, 45, 50 weight %) (TBCHA, EGDMA, THFA, HPMA), adhesion promoter (2,5 weight %) (Ebecryl 1291), photoinitiator (2 weight %) (Lucirin TPO_L), and base oligomer (rest amount) (Ebecryl 1258) - were mixed (100-300rpm) at room temperature (23°C) in yellow light (589nm) for 10 min. Uncured gels were stored in a thermostatically controlled environment T=40°C for 24h to remove the trapped air. The mixed gels were used for viscosity/flowability tests or were applied on the substrate (Teflon or aluminium sheets) in thickness 200 µm and cured for 30s under UV lamp (*KP800LED*) with luminous intensity 130mw and spectrum maximum λ = 405nm (*YI Liang electron technology Co, China*) yielding cross-linked films for mechanical properties, microhardness, surface gloss and adhesion loss tests.

2.3. Test methods

The effect monomer concentration on the viscosity of urethane acrylate mixture was determined by Brookfield viscometer *DV-II + Pro (Brookfield Engineering Laboratories*). The systems with a viscosity in the range from 1000 up to 6000mPas were searched.
The degree of conversion was determined by evaluation of gel fraction of the cured coating. To evaluate post curing effects gel fraction of the coatings was determined immediately after cure (GF) and after 72 hours of storage (GF_0). Samples were inserted in a Soxlet apparatus and maintained in a condensing acetone vapor atmosphere for 48 hours. Residual gel was dried in vacuum to constant weight. GF was calculated as follows (1), were m_o is the initial sample's weight and m_t is the mass of tested samples after drying.

$$\%GF = \left[\frac{m_t}{m_0}\right] \times 100(1)$$

The mechanical properties of UV-cured films were determined by standard tensile stress-strain tests to measure tensile strength (σ) and elongation (ε) immediately after cure and after 72h of storage. Standard tensile stress-strain experiments were performed at room temperature using *BDO FB020TN (Zwick Roell, Germany)* universal testing machine according to the requirements of *LVS EN ISO 527-3:2000* [20].
Gloss measurements of the coatings immediately after cure (G) and after 72 hours of storage (G_0) were carried out using the *Novo Gloss Lite Gloss Meter (Rhopoint Instruments Ltd.*) on a black substrate at an angle of 60°.
Micro hardness of the coatings immediately after cure (H) and after 72 hours of storage (H_0) experiments were performed using Vickers microscope M41 at 200g statistic loading.
Adhesion loss of the coatings to the substrate was evaluated by a following procedure.. Initially 200 µm thick 5cm x 5cm cured films on aluminum sheets were left for 72 hours to assure complete polymerization and afterwards removability of the coating was evaluated. Coatings were wrapped in paper, impregnated in 95% acetone solution with water, coated with aluminum foil (100 µm thick) and kept for 10 minutes. Consequently scratch resistance, an indicator of coating adhesion to the substrate, was measured with SAM (Scratch Adhesion and Mar Tester) device *S.A.M PA-5050* (*Gardco*) where the Hoffman tester concept is used [21]. After each 30 second period, coating adhesion to the substrate was evaluated using 5 SAM scratches (with distance 5mm, force 250g), horizontally and vertically. When 100% adhesion loss between the substrate and the coating appeared, effective soak-off time was detected.

3. Results and discussions

Natural nail has a complicated porous structure and there is no close material to natural nail to test the adhesion. Fingernails are composite structures containing fibrils and plate-like structures of keratin bound together by a keratinous matrix material [23]. Based on previous experience [18, 19] we put forward the hypothesis that adhesion is dependent on the degree of conversion of the coating. Our assumption is that the lower the degree of conversion – the faster should be the adhesion loss from the substrate. At the same time degree of conversion should be enough high to provide the required mechanical characteristics: elasticity, toughness, surface gloss and adhesion in the time of use. Also degree of conversion cannot be lower than 90% due to the health issues, where residual monomers should not be found in the cured coating [24].

3.1. Viscosity

For the coating application and other processing activities, viscosity is the most important parameter. It should be in the range of 3000-4500 mPas to provide the good leveling, but not smaller, to avoid flowing. To select the optimum monomer concentrations this range has been expanded from 1000 to 6000mPas. To find the optimal concentrations, first 8 different concentrations were tested and results are shown in Figure 1.

All four monomers ensure the required viscosity by considering that their concentrations in the formulation are 30%, 35% and 40%. Formulations with 30% and 40% concentrations of the monomers have been chosen to evaluate the influence of the monomer on the exploitation characteristics of the cured coatings.
Nail coating application requirements (viscosity 3000-4500mPas) have been met for the compositions containing 30% EGDMA, 40% HPMA and 30-40% TBCHA. Further investigations for aged uncured samples are necessary to predict the storage time for uncured compositions.

3.2. Degree of Conversion
In this test eight different samples were prepared and characterized. Samples differed with the monomers included and weight ratio between oligomers and monomers: 70:30 and 60:40. Gel contents of polymeric films were determined immediately after cure (GF) and after 72h of storage (GF_0). Results were found to be between 94 and 99% as it is shown in Table 1.
Degree of conversion of monomer EGDMA was the highest, providing the GF between 96.4% and 98.5%. Such behavior was expected due to the higher functionality of the acrylate monomer.

Table 1. Monomer effect on the gel fraction (GF), surface gloss (G) and micro hardness (HV) of the cross-linked films.

	Monomer	Conc.	Gel fraction			Surface gloss			Micro hardness		
			GF	GF_0	$\Delta GF^*/GF_0$ x 100	G	G_0	$\Delta G^{**}/G_0$ x 100	HV	HV_0	$\Delta HV^{***}/HV_0$ x 100
		%	%	%	%	GU	GU	%	MPa	MPa	%
1	TBCHA	30	97,4	98,0	0,61	46	**68**	32	104	**144**	38
2	TBCHA	40	94,0	**94,1**	0,11	51	**72**	29	69	83	20
3	EGDMA	30	98,5	99,3	0,81	47	59	20	204	**281**	38
4	EGDMA	40	96,4	98,9	2,53	47	57	18	220	**314**	43
5	THFA	30	96,0	96,3	0,31	59	**73**	19	25	47	88
6	THFA	40	95,7	96,7	1,03	54	61	11	28	32	14
7	HPMA	30	97,0	97,6	0,61	50	63	21	99	**151**	53
8	HPMA	40	95,0	99,9	4,90	43	58	26	65	**134**	106

* $\Delta GF = (GF_0 - GF)$ ** $\Delta G = (G_0 - G)$ *** $\Delta HV = (HV_0 - HV)$
****Characteristics in bold are in requested range or close to that

Gel contents of all the investigated systems increased in time by 0.1% - 4.9% (after 72 hours of storage) due to post-curing. The extracted uncured parts of 4 different kinds of monomers increase with increasing content of the monomer (40%).
In other studies it has been found that microstructure, crosslink density as well as degree of conversion are strongly related to the mechanical and other properties of the materials [6, 23]. In this study we found correlation between degree of conversion not only with mechanical properties, but also with adhesion.

3.3. Mechanical properties
Young's modulus of natural nail at relative humidity 55% is 2.32 GPa [24]. Deformation characteristics of the coating on the nail should have as close as possible to those of the natural nail to ensure good adhesion and avoid air pockets. No one of the tested compositions reached so high modulus of elasticity. Films containing bi-functional monomer EGDMA reached the highest E values, but cured films were brittle and with the smallest ultimate relative deformation ε_B values (Fig.2.).
Stability of mechanical properties in time is very important for natural nail coatings. Modulus of elasticity increases for all studied compositions along with storage time, whereas the highest increase -

57% - have been found for the composition containing 30% THFA, and for both concentrations with HPMA (37% and 54%). Systems containing TBCHA showed the most stable E values; post-cure time have almost no effect on the characteristic of the cured films.

As natural nail coating should have the ability to move together with natural nail plate, coating should have ε_B at least 0,6% [22]. Initial tests showed that ε_B should be in range 20-50%, to avoid too high brittleness or the softness of the film. Compositions with TBCHA and HPMA fit in the requested range. These compositions should be tested more carefully to detect their long term adhesion, removability, surface hardness and surface gloss.

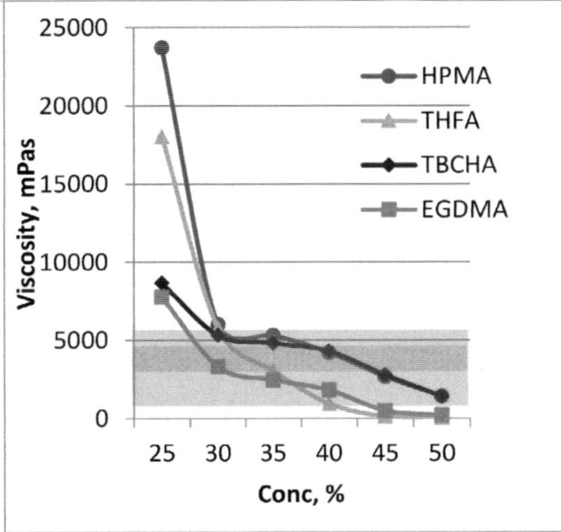

Figure 1. Effect of monomer concentration on the viscosity of uncured system in requested viscosity range

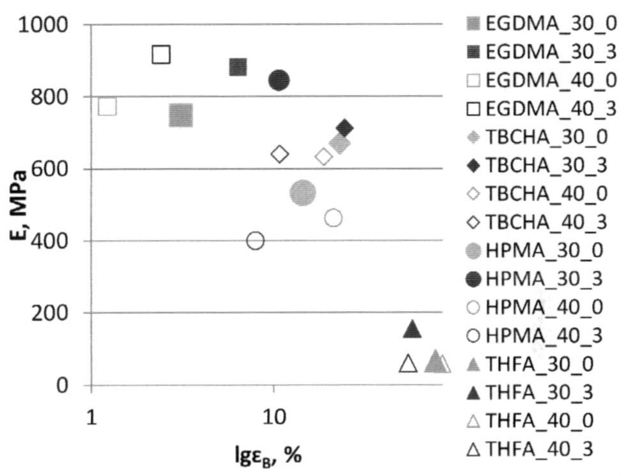

Figure 2. Elastic modulus and elongation at break values for the films with 30% and 40% of monomers measured immediately after cure (0) and after 3 days of storage (3).

3.4. Gloss

The effect of the monomer type and concentration on the gloss were investigated using the constant curing conditions (time, intensity, layer thickness) and constant photoinitiator amount. Used monomer affected the final gloss of the samples in the range from 43 to 73 GU (Table 1.). Monomer concentration increase in the system differently affects the gloss values. Concentration of EGDMA doesn't affect the gloss, TBCHA increases it, but other 2 lower molecular weight monomers (HMPA and THFA) decrease it.

For all the coatings the surface gloss values increase in the time due to the post cure effect. Values increase by 15-35%. For top coat the gloss values are very important characteristic to ensure good esthetic look of the coating. Thus compositions with 30% or 40% of TBCHA are recommended especially for top coats.

3.5. Adhesion loss

Coating stability is one of the most important properties for the chosen application. All previous studies have been done mostly on long term adhesion properties, but this application requires both: good adhesion in the time of functionality and fast adhesion loss at coating removal. Soak–off tests have been performed after 72h of storage and results are presented in the figure 3. Obtained results correlate with the GF data obtained after 72h of storage. The greatest adhesion loss was obtained for the composition containing 40% TBCHA for which adhesion from substrate was lost after 3 min and it correlates with the smallest GF values.

Compositions with EGDMA took the longest time for the removal. This monomer has the closest mechanical properties to the natural nail, but ensures too strong adhesion with natural nail, due to its higher reactivity (bi-functionality). TBCHA shows average mechanical characteristics and good adhesion loss ability; therefore compositions with 30% and 40% of TBCHA should be developed further to improve mechanical properties.

3.6. The composition influence on coatings characteristics

Short term coatings challenges with, at least, seven characteristics, which should meet the requirements: viscosity, efficient polymerization, gloss, surface hardness, toughness, elasticity, removability. All ingredients in the composition affect the final product properties.

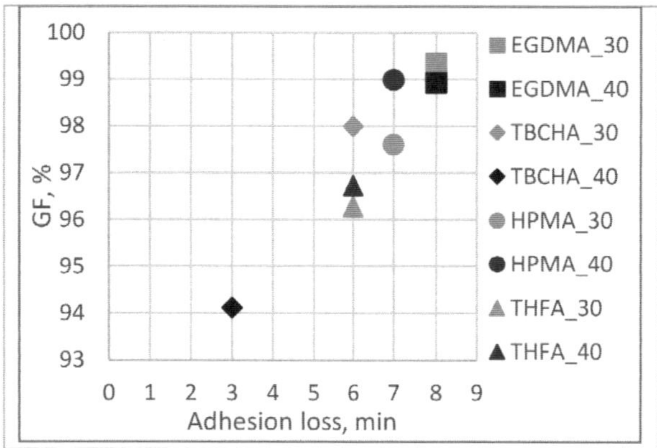

Figure 3. Correlation of gel fraction data (after 72h of storage) of the compositions with adhesion loss time from the substrate in a solvent medium

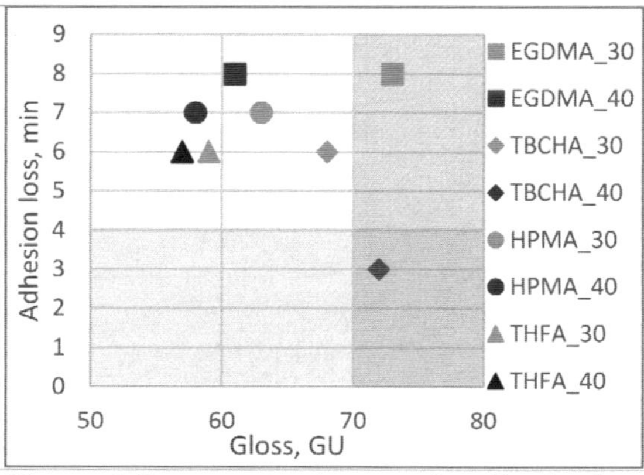

Figure 4. Requested value area for adhesion loss time (0-4 min.) and gloss (70-80 GU); and composition characteristic (after 72h of storage) location in them

In this study we evaluated the monomer effect on the coating characteristics, but no one of the monomers have been found to be perfect and could improve all the characteristics. The results obtained are presented in figure 4, where requested values for adhesion loss (0-4 min.) and gloss (70-80 GU) are compared. As shown, only one composition (with 40% TBCHA) fits with requested levels of both of the parameters. Although the compositions with 30% of TBCHA show the results quite close to the requested, their film toughness should be increased.

4. Conclusions

Assumption, that lower degree of conversion promotes the destruction of adhesion between the substrate and the coating, was confirmed. Presence of 40% of the monomer TBCHA in the composition showed the smallest GF value (94%), what provided the fastest adhesion loss.

Good coating flexibility and consistency of mechanical properties can be achieved by using 30% of the monomer of TBCHA in the composition, which showed the optimal ε_B values (23.07-24.58%), high E values (670-712MPa) and comparatively constant properties in 72 hours (ΔE =1.25%, $\Delta\varepsilon_B$=6.14%). A composition with 40% of TBCHA showed the fastest coating destruction times, achieving adhesion loss from the substrate after 3 min.

Acknowledgements

The authors thank the Investment and Development Agency of Latvia (Project No. LKC/2.1.2.1.1/10/01/006, Contract No. L-KC-11-0005) for their financial support.

IOP Conf. Series: Materials Science and Engineering **111** (2016) 012008 doi:10.1088/1757-899X/111/1/012008

References

[1] Liu P, Gu A, Liang G, Guan Q andYuan L. 2012 *Prog.Org.Coat***74** 150
[2] Sauro S, Vijay S and Deb S2012 *Eur.Polym.J***48** 1474
[3] Milinavičiūtė A,Jankauskaitė V and Narmontas P 2011*Mater. Sci-medzg***17**216
[4] Jančovičová V, Mikula M, Havlínová B and Jakubíková Z 2013*Prog.Org.Coat***76**438
[5] Wang F, Hu J.Q and Tu W.P2008 *Prog.Org.Coat***62**250
[6] Kayaman-Apohan N, Demirci R, Cakir M and Gungor A 2005*Prog.Org.Coat***73**262
[7] Pichavant L and Coqueret X. 2008 *Prog.Org.Coat*62
[8] Łopaciuk A and Łoboda M. Global Beauty Industry Trends in the 21st Century Managemant, Knowledge and Learning International Conference 2013 Zarad.
[9] Drummey C. 2013-2014 Industry Statistics Highlights Nails Magazine 2013
[10] Vu T and Chad C. Compositions and Methods for Nail Coatings, United States Patent 20110182838, 28 July 2011
[11] Haile D.L Compositions for Removable Gel Applications for Nails and Methods of their Use, USA Patent WO 2011/011304 A2, 27 January 2011
[12] Koschar M. Easily Removable Nail Polish Composition, Germany Patent DE102011102661 A1, 29 November 2012
[13] Wu N, Zhang Y and Wang Y 2013*Int. J. Adhes*
[14] Oguri M, Yoshida Y, Yoshihara K, Miyauchi T, Nakamura Y, and Shimoda S 2012 *Acta Biomater*
[15] Misbah S et al. 2012 *Carbohyd. Polym* **87** 404
[16] Gällstedt M, Törnqvist J and Hedenqvist M.S *2001 J. Polymer Sci. Part B: Polymer Phys***39**992
[17] El-Shafee et al.2001 *Eur. Polymer J***37**2104
[18] Grigale-Sorocina Z,Kalnins M et al. *Mater. Sci.* (Medziagotyra) in print.
[19] Grigale-Sorocina Z, Kalnins M et al. 2015 *P. Est. Acad. Sci***64** 93
[20] Determination of tensile properties. Part 3: Films and sheets, test conditions (in Latvian), BS EN ISO 527-3: 2000th Plastic.
[21] Koleske J. V 1995*Paint and Coating Testing Manual.* American Society for Testing and Materials, Philadelphia
[22] Farran L. et.al. 2009 *J Biomech***42** 1235
[23] Taki K and Nakamura T. J*Cosmet., Dermatol. Sci. Appl*
[24] Alishiri M, et al. 2014 *Mater. Sci. Eng***42** 773

Poly(vinyl alcohol) – poly(vinyl acetate) composite films from water systems: formation, strength-deformation characteristics, fracture

N Geisari, M Kalnins
Riga Technical University, Faculty of Material Science and Applied Chemistry, Institute of Polymer Materials, P. Valdena Street 3, Riga, LV-1048, Latvia

E-mail: natalja.jelinska@rtu.lv

Abstract. The object of investigation was thin (< 200 μm) films, formed by casting from water based mixtures of poly(vinyl alcohol) (PVA) water solution and poly(vinyl acetate) (PVAc) water suspension. Starting systems hence were suspensions of plasticized PVAc particles in PVA water solution. Systems with full range of PVAc volume fraction φ_{PVAc} were studied. Kinetic relationships of several parameters which characterize film formation in drying process were compared and scrutinized. It revealed that the main factor which restrains coalescence of PVAc particles in the composite is high viscosity of PVA water solution at the final stage of composite structure formation. In tensile deformation conditions PVA/PVAc systems behave as being PVA matrix composites in a whole range of PVAc content. Use of *Kerner* approach, based on the self-consistency idea, for the analysis of $E_{COMP}(\varphi_{PVAc})$ relation, confirms the continuity of PVA matrix.

1. Introduction
Poly(vinyl alcohol) (PVA) is the most widely produced water soluble polymer today. PVA is a polymer of a great interest because of its many desirable characteristics: high mechanical properties, biodegradability, excellent gas barrier properties, transparency. Creating of blend systems with other polymers is promising way of modification of strength-deformation properties of PVA. Blends of PVA with different linear polymers were investigated for different purposes [1-4].

In the present study, blends of PVA with the plasticized poly(vinyl acetate) (PVAc) were investigated. PVA/PVAc film specimens were manufactured by casting blends of a PVA water solution and a PVAc water suspension. At room temperature the components form separate phases [5]. The purpose of the present study is to estimate the structure and its correlation with certain strength- deformation characteristics of PVA/PVAc blend films.

2. Materials and Investigation Methods
PVA with a molecular weight of 133,000 Da and a content of acetate groups 0.98%, as well as a PVAc (containing 10% of a dioctyl phthalate as plasticizer) water suspension were used as initial materials. A water solution of PVA was mixed with a PVAc suspension (10% of polymer in both cases) in different proportions providing the specified volume fraction of the components (φ_{PVA} and φ_{PVAc}) in the blend. The films were obtained by casting: the blends were poured into *Petri* dishes 140 mm in diameter (the initial thickness of the layer was 2mm) and dried at 17°C up to a constant content of water (≈ 3%).The thickness of obtained films was 150 ± 10 μm. Water content W_{H20} in the drying system were determined by continuous weighing. Viscosity μ of PVA water solution as function of W_{H20} was determined using Thermo HAAKE VT 550 rotary viscometer at 25°C with cylinders SC4-18, SC4-34, ULA-34.

The films were tested in tension on a Zwick/Roell universal machine at room temperature according to ENISO527. The speed of the mobile crosshead was 20mm/min. The data obtained were processed by using the test XpertV11.0 program. The stress–strain relationship σ–ε was determined up

IOP Conf. Series: Materials Science and Engineering **111** (2016) 012009 doi:10.1088/1757-899X/111/1/012009

to the failure of specimens. All mechanical characteristics were found on the basis of five parallel measurements.

SEM was used to determine morphology of PVA/PVAc systems. The specimens were cooled in liquid nitrogen and subjected to a brittle fracture. The fracture surfaces were gold covered (Emitech,K550X). Samples were examined on a Tescan, Mira//LMU Schottky, scanning electron microscope at 3000× and 10,000× magnifications and a voltage of 25 kV.

3. Discussion of the Results

Change in time of the most important parameters of the drying PVA/PVAc (for $\varphi_{PVAc} = 0,5$) layer withis shown in figure1 as an example.

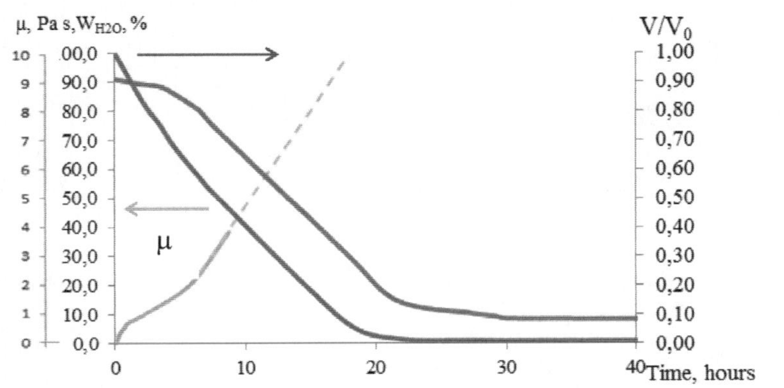

Figure 1. Change of the water content W_{H20} in PVA aqueous solution, volume of PVA/PVAc system V (V_0 - initial volume of the system) and viscosity μ of PVA solution of PVA/PVAc system with time t ($\varphi_{\text{PVAc}} = 0,5$) in drying process at room temperature.

In a course of drying the water content in PVA solution decreases, content of PVAc particles in the suspension grows, the distance between particles decreases. The greater is the volume fraction of PVAc (φ_{PVAc}) in PVA/PVAc system the smaller the final average distance between the PVAc particles. Up to $\varphi_{PVAc} < 0.7$ plasticized PVAc particles are more or less evenly dispersed in PVA medium. Associates PVAc particles did not observed. Evidently closer merger of particles is hindered by large viscosity of PVA solution. Particles retain the original spherical shape (figure 2 a - e).

The situation changes when φ_{PVAc}value is greater. The same complex of interfacial and capillary forces,which determines the PVAc film formation from pure suspension, begins to play a decisive role. The difference is that instead of water there is PVA water solution, which viscosity and surface tension are changing during the drying process. PVAc particles are gradually deformed while moving towards each other, resulting in a form that is close to polyhedron. Thickness of PVA interlayer decreases (see figure 2 e,f).

From figure 2 f it is seen that direct contact particles and coalescence of PVAc does not happen, even at φ_{PVAc}= 0,9 and the PVA interlayer between PVAc particles remains. SEM micrographs of PVA/PVAc system with φ_{PVAc}= 0,9 (figure 2 f) and pure PVAc (figure 2 g) shows that the morphology of corresponding systems differs. In the system with φ_{PVAc}= 0,9 PVA particles are less compactly arranged and PVA interlayer persists.

The values of tensile elastic modulus E of composites are between the E values of rigid PVA (E_{PVA} = 2,3 MPa) and compliant PVAc (E_{PVAc} = 0,3 MPa). E value decreases with an increase of φ_{PVAc}: namely, increase of content of compliant component in the rigid matrix (figure 3).

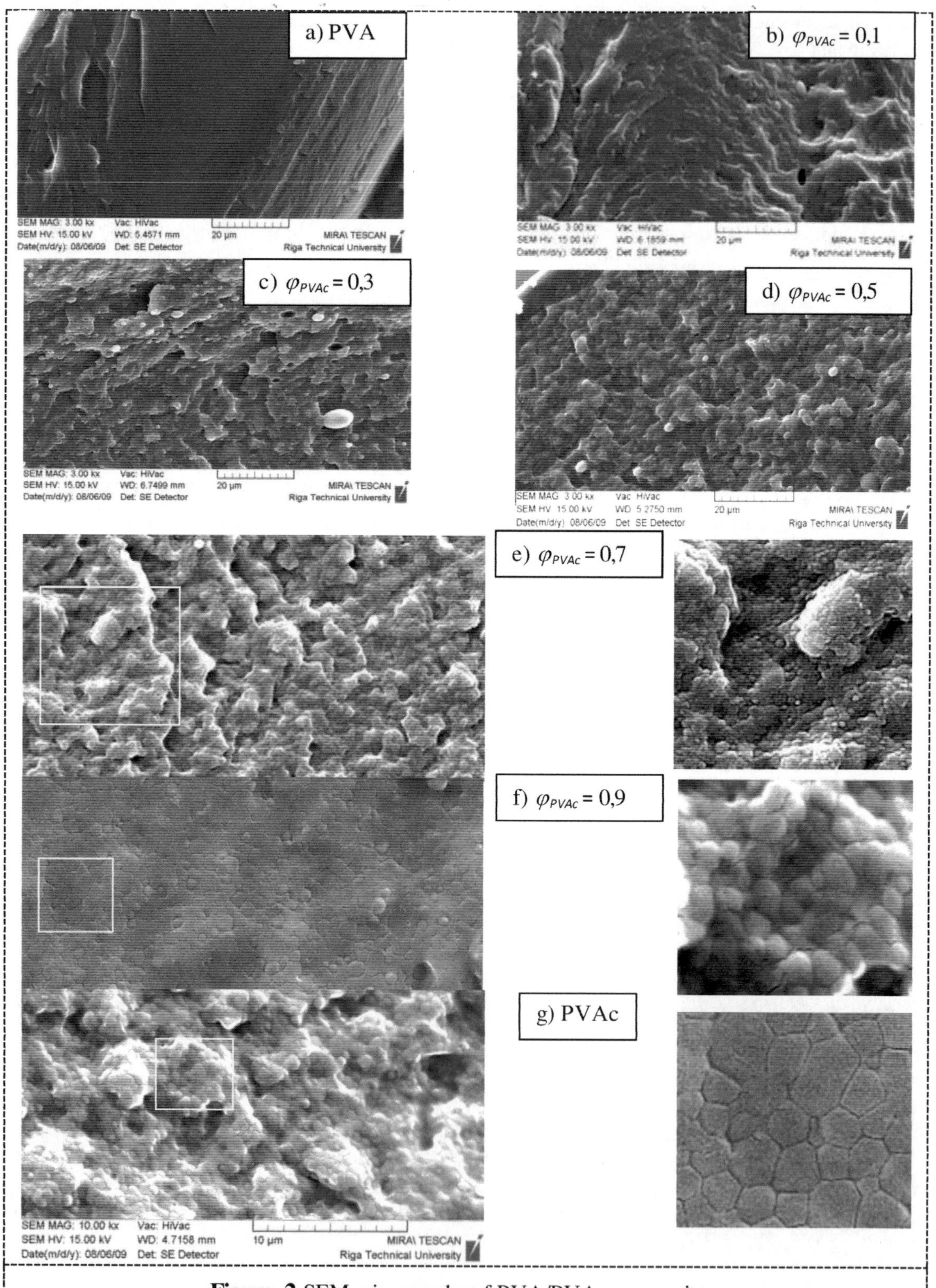

Figure 2.SEM micrographs of PVA/PVAc composites.

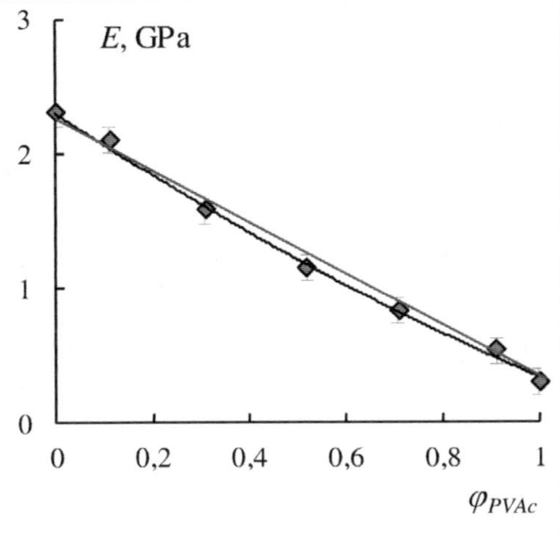

Figure 3. Values of modulus of elasticity E of PVA/PVAc blends depending on PVAc content in composite φ_{PVAc}.

The dependence of composite tensile elastic modulus E on φ_{PVAc} is close to additive one (red straight line):

$$E = E_{PVA} - (E_{PVA} - E_{PVAc})\, \varphi_{PVAc}. \tag{1}$$

Thereby from the point of view of rigidity the studied composite essentially is rigid PVA matrix in which more compliant PVAc particles are included.

To confirm the continuity of PVA matrix *Kerner* approach (based on the self-consistency idea [6]), was used for the analysis of $E(\varphi_{PVAc})$ relation.

The value of modulus of elasticity E* of composite in accordance with *Kerner* model is calculated as follows:

$$E^* = \frac{9KG}{3K+G} \tag{2}$$

Where: K – calculated value of bulk elasticity modulus and G – calculated value of shear elasticity modulus of composite.

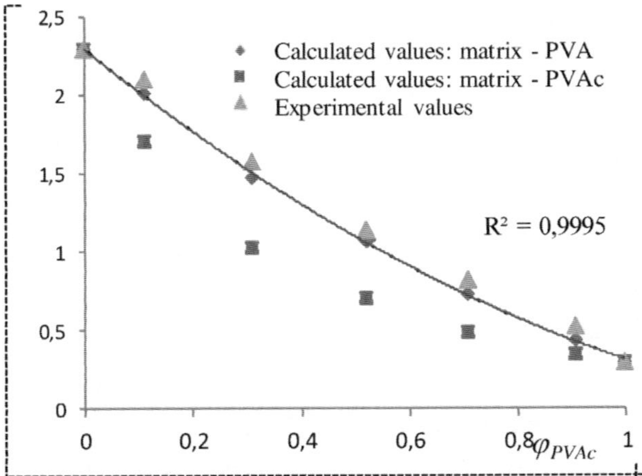

Figure 4. Calculated values (*Kerner* model) of modulus of elasticity E^* compared with the experimentally determined modulus of elasticity E values at different PVAc content in the composite (φ_{PVAc}).

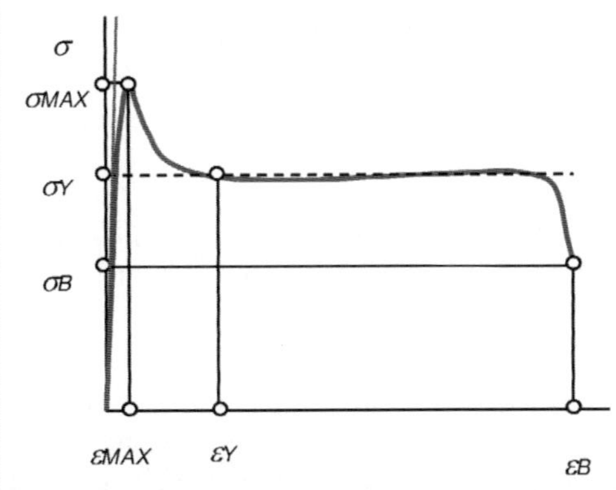

Figure 5. Schematic representation of σ(ε) curve of PVA/PVAc composite and obtainable characteristics: ($E = lim\ (d\sigma/d\varepsilon)|_{\varepsilon\to 0}$).

Calculation of K, G and E^* values were performed in two ways: either assuming that one of the components is a "matrix"(index-m) and the other is "filler", included in the matrix (index - f) or vice versa (components alternate status).

K and G values of composite were calculated as functions of K_f, K_m, G_m and φ_f (volume fraction of „filler"). K and G values of PVA and PVAc were calculated as follows:

$$K_{PVAc}= E_{PVAc}/3(1 - 2\ \nu_{PVAc}) \tag{3}$$

$$K_{PVA}= E_{PVA}/3(1 - 2\ \nu_{PVA}) \tag{4}$$

$$G_{PVAc}= E_{PVAc}/2(1 + \nu_{PVAc}) \tag{5}$$

$$G_{PVA}= E_{PVA}/2(1 + \nu_{PVA}) \tag{6}$$

Where: E_{PVAc} un E_{PVA} – accordingly experimental values of tensile elastic modulus of PVAc and PVA: E_{PVAc} = 0,3GPa, E_{PVA} = 2,3GPa. ν_{PVA} un ν_{PVS} – accordingly PVA and PVS Poisson coefficient values: ν_{PVAc} =0,16 [7] and ν_{PVS} =0,44.

Experimental values of modulus of elasticity are very close to the calculated values in the case where the matrix is PVA (see figure 4). This is an serious evidence, that PVA remains as a continuous composite matrix in a wide range of PVA content (at least at φ_{PVAc}> 0,1).

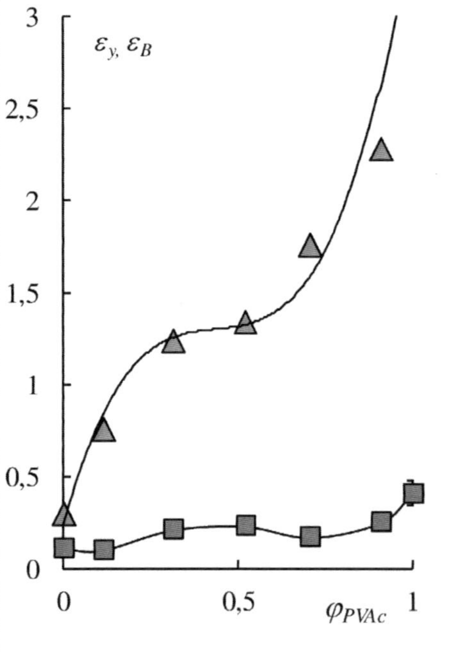

Figure 6. Dependence of relative elongation values on composite content.

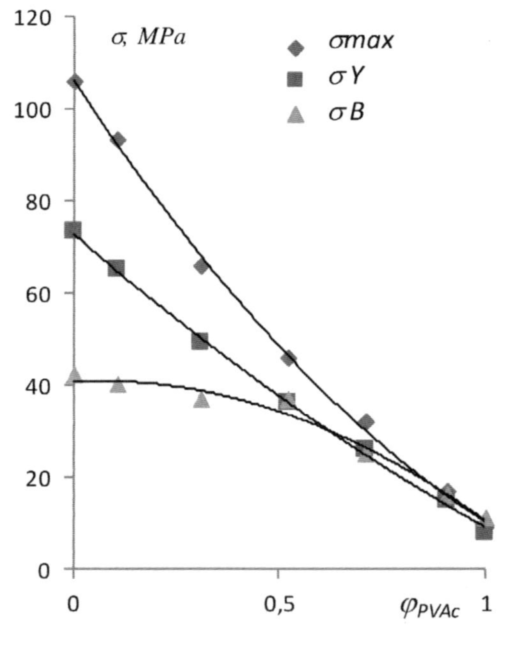

Figure 7. Dependence of σ_{MAX}, σ_Y un σ_B on volume fraction of PVAc in composite.

Deforming the composite over ε_{MAX} and yield elongation ε_Y (see figure 5) the PVA frame gradually locally crashes at stress values between σ_{MAX} (maximum stress,) and σ_Y (yield stress, figure 7). It occurs in locations where the deformation value exceeds the PVA relative elongation at break ε_{BPVA}= 0,29. PVAc particles start to deform (relative elongation at break of PVAc is more than ten times higher ε_{BPVAc} = 3,6) (see figure 6).

The σ_{MAX} and σ_Y values are determined by PVA matrix frame. The lower volume fraction of PVA matrix (φ_{PVA}) in the composite (less effective load-carrying part of frame), the lower are σ_{MAX} and σ_Y values. At higher ε values composite loses integrity and collapses at comparatively love stress value (σ_B), either by fracture of PVAc particles or by flinch them from PVA matrix. Obtained results suggest slight change in the data interpretation of paper [8].

4. Conclusions

Up to PVA volume fraction 0,7 PVA/PVAc composite consists of PVA matrix in which not deformed PVAc particles more or less evenly are distributed. At higher content of PVAc interfacial forces deforms spherical particles close to polyhedron. The main factor which restrains coalescence of PVAc particles is high viscosity of PVA water solution at the final stage of composite structure formation.

PVA remains as a composite matrix in a wide range of PVA content (φ_{PVAc}> 0,1). The dependence of composite tensile elastic modulus E_{COMP} on φ_{PVAc} is very close to additive one: $E_{COMP} = E_{PVA} - (E_{PVA} - E_{PVAc})\ \varphi_{PVAc}$. Composite essentially is a rigid PVA frame in which much more compliant PVAc particles are included.

Correlation analysis of characteristics, obtained from stress–strain relations for composites with various PVAc content shows that fracture of composites starts with break of less deformable load-bearing PVA frame at comparatively small elongation and continues with significant deformation of PVAc component. This feature becomes more pronounced with increase of PVAc content.

References:

[1] Peppas NA and Tennenhouse N 2004 Semicrystalline poly(vinyl alcohol) films and their blends with poly (acrylic acid) and poly (ethylene glycol) for drug delivery applications *J. Drug Del. Sci. Tech.* **14**(4) 291-297

[2] Vasile C and Kulshreshtha AK 2003 *Handbook of Polymer Blends and Composites* **4** (Rapra Technology)

[3] Patachia S, Vasile C and Mavru E 1985 Study of the compatibility in the poly (vinyl chloride)/poly (vinyl alcohol) *Polymer Bulletin* **13**(4) 301-306

[4] Chun-Chen Y, Yang JM and Cheng-Yeou W 2009 Poly (vinyl alcohol)/poly (vinyl chloride) composite polymer membranes for secondary zinc electrodes *J. Power Sources* **191**(2) 669-677

[5] Jelinska N, Kalnins M, Tupureina V and Dzene A 2010 Poly (vinyl alcohol)/poly (vinyl acetate) blend films *RTU Mater. Sci. Appl. Chem.* **1**(21) 55-61

[6] Kerner EH 1956 The elastic and thermoelastic properties of composite media *Proc. Phys. Soc.* **69B** 808-813

[7] Cui HW and Du GB 2011 Preparation and characterization of exfoliated nano-composite of poly vinyl acetate and montmorillonite *J. Chem. Eng. Mater. Sci.* **2**(8) 122-128

[8] Jelinska N and Kalnins M 2011 Strenght and deformation characteristics of polymer blend films obtained from water systems *Mech. Comp. Mater.* **47**(5) 236-250

IOP Conf. Series: Materials Science and Engineering **111** (2016) 012010 doi:10.1088/1757-899X/111/1/012010

Delamination Detection in Carbon Fibre Reinforced Composites Using Electrical Resistance Measurement

A Kovalovs[1], S Rucevskis[1], V Kulakov[2], A Aniskevich[2]
[1]Riga Technical University, Institute of Materials and Structures, Azenes Str. 16, LV-1048, Riga, Latvia
[2]University of Latvia, Institute of Polymer Mechanics, Aizkraukles Str. 23, LV-1006, Riga, Latvia

E-mail: andrejs.kovalovs@rtu.lv

Abstract. In the present study 2-D numerical analysis of strip-type laminated composite specimens with and without damage is considered and numerical investigation is carried out by using a finite element method. The surface and oblique resistances are numerically calculated according to the two-probe and four-probe methods. The electrical conductivity of the composite laminate in the longitudinal direction is constant, while the electrical conductivity in the through-thickness direction is used as a variable in the parametric study. The resistance change due to delamination for each case is estimated by comparing the obtained resistance with the corresponding resistance of the specimen without delamination. Applicability and effectiveness of the proposed method are investigated by using various lengths of a delaminated crack in the specimen.

1. Introduction

Composite materials, such as carbon-fibre-reinforced polymers, are extensively used in the fabrication of high-performance structures for a variety of engineering applications such as aircraft, automobiles, sporting goods and electronics. Composite laminates often show considerable advantages of stiffness and strength over homogeneous materials. However, these advantages are counterbalanced by lower damage tolerance level. External mechanical loads, repeated cyclic stresses, and impact lead to damage in laminated composites due to their low delamination resistance. Damage in a structure may cause failure leading to tragic consequences and therefore structural health monitoring and damage detection in civil, mechanical and aerospace engineering structures has become one of the most important keys in maintaining the integrity and safety of a structure.

Several methods have been developed for the structural health monitoring of composite structures using fibre optic sensors [1], piezoelectric sensors [2], and electrical resistance change method [3-10]. The electrical resistance change method has been employed by many researchers for detection of internal delamination. This method does not require expensive instruments, and it is applicable to existing structures. The basic idea of this method is that damage such as fibre fracture or delamination between plies will cause a decrease of the electrical conductivity in the damaged region leading to a resistance or voltage change. For the measurement of electrical properties of laminated composites the two-probe and four-probe methods are employed.

The aim of this study is to investigate electrical resistance changes of strip-type laminated composite specimens with and without delamination by using the finite element method (FEM). The electrical resistance change method for delamination detection based on the electrical resistance

IOP Conf. Series: Materials Science and Engineering **111** (2016) 012010 doi:10.1088/1757-899X/111/1/012010

measurement of laminated composites is proposed. The surface and oblique resistances are numerically calculated by using the two-probe and four-probe methods. Applicability and effectiveness of the proposed methods are investigated by using various lengths of a delaminated crack in the specimen.

2. The two-probe and four-probe methods

The techniques for the measurement of the electrical behaviour of laminated composites are the two-probe method and four probe methods. The two-probe method is based on the definition of resistance when two electrodes are used to measure the electrical resistance. Figure 1 schematically shows a specimen, on which a pair of contacts (probe 1 and probe 2) with conductive wires is attached. Two probes are used for the electrical current input, as well as for the voltage measure. The resistance of the segment between the voltage contacts can be calculated through Ohm's law:

$$R = \frac{V}{I} \tag{1}$$

where, V and I are the voltage and current from the voltage and current contacts, respectively.

The four-probe method is an alternative to the two-probe method. Electrical current is passed through the outer probes and induces voltage in the inner probes. Based on the measured voltage and current, the resistance between the voltage contacts is then calculated. The four-probe method is schematically shown in Figure 2.

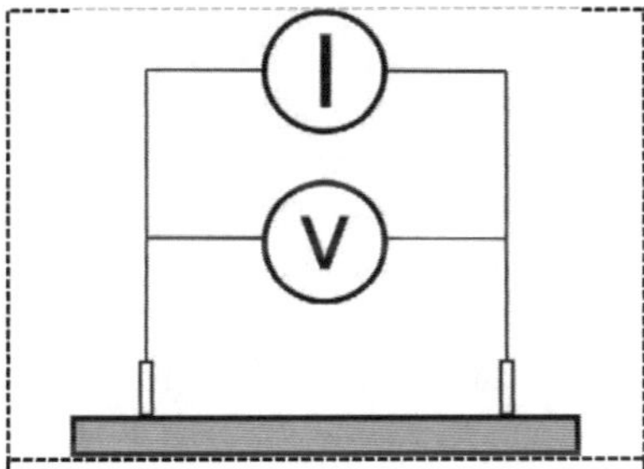

Figure 1. Two probe method.

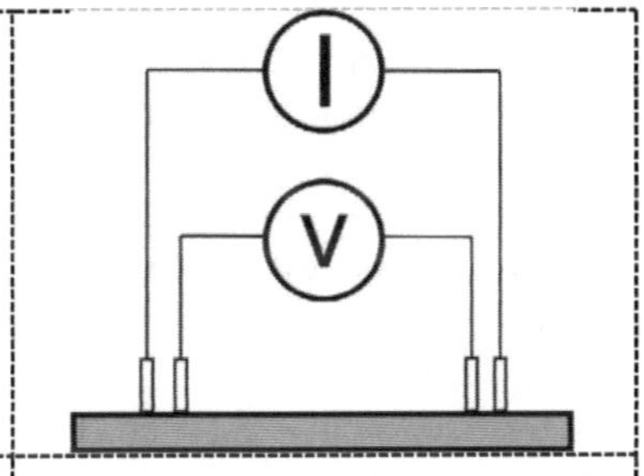

Figure 2. Four probe method

3. Numerical analysis

A 2-D strip-type laminated composite specimen is designed for FEM analysis of the delamination monitoring, as shown in Figure 3. The dimensions are similar to the specimen used by Wang et al. [4, 5], i.e., L_0 = 200mm, L_1 = 10mm, L_2 = 30 mm, L_3 = 2 mm and H =3.2 mm. Figure 4 shows the location of delamination. The specimen has seven equally spaced electrodes mounted on the top and bottom to measure the surface and oblique resistance by the two- and four-probe methods.

The electrical conductivity in the longitudinal direction is taken as σ_l=15Ω^{-1}mm^{-1} and the electrical conductivity in the through-thickness direction σ_t is used as a parameter to carry out a parametric study. For a composite panel the through-thickness conductivity σ_t is usually much smaller compared to the longitudinal conductivity σ_l. The conductivity of the electrodes is taken as 10^{8}Ω^{-1}mm^{-1} and in this case the electrode resistances can be assumed negligible.

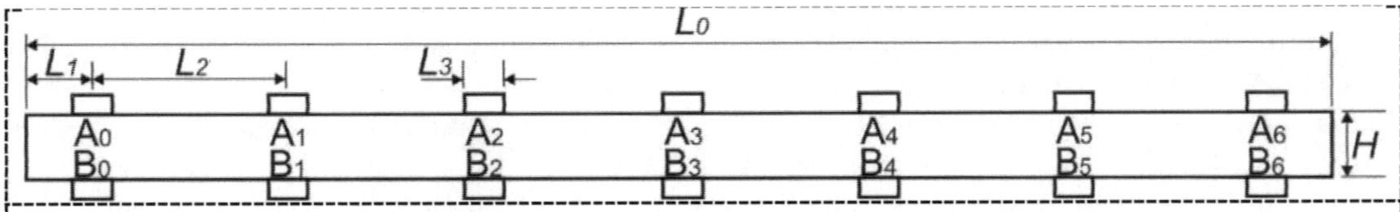

Figure 3. 2-D specimen and electrode locations.

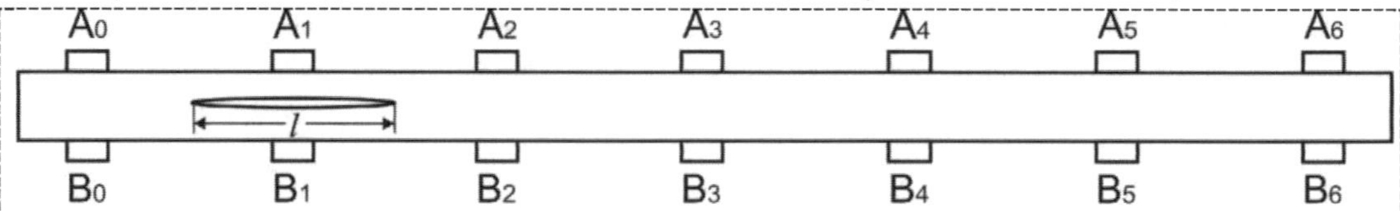

Figure 4. Delamination location between electrodes A_1 and B_1.

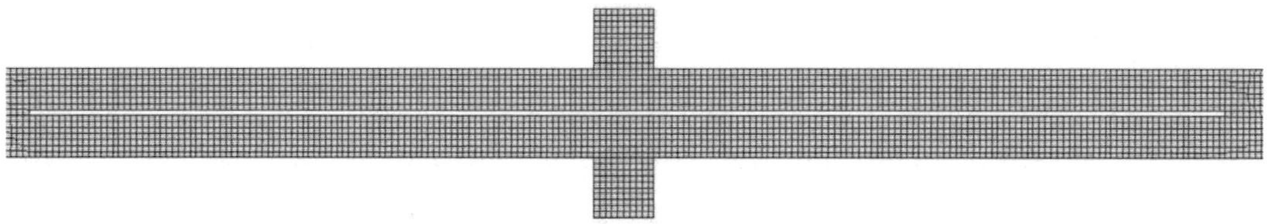

Figure 5. Fragment of the finite element model with delamination.

A 2-D finite element model of the composite strip-type specimen is designed by using the finite element method code ANSYS. Figure 5 shows the fragment of the finite element mesh for a specimen with delamination. A delamination crack is created by cutting off elements from the finite element model.

The electrical resistances between electrodes located on the same side of the specimen and electrodes located on different sides of the specimen are named as the surface and the oblique resistance, respectively. The resistances are obtained by using the two- and four probe methods. To compute the surface resistance R_{A1A5} by the two-probe method, a direct electric current of 10 mA is charged from electrode A_1 to electrode A_5. The electrical voltage of electrode A_5 is set to 0 V. Electrical voltage obtained at the electrode A_1 is divided by the electrical current (10 mA) to calculate the electrical resistance of the segment A_1-A_5. To compute the surface resistance R_{A1A5} by the four-probe method, a direct electric current of 10 mA is charged from electrode A_0 to electrode A_6 and the electrical voltage of the electrode A_5 is set to 0 V. Electrical voltage obtained at the electrode A_1 is divided by the electrical current (10 mA) to calculate the electrical resistance of segment A_1-A_5. For the oblique resistance measurements, instead, a unit current I is applied as input to the electrode pair A_1-B_5 or A_0-B_6, and the voltage V between the electrode pair A_1-B_5 as output is obtained from numerical experiments.

4. Results and discussion

4.1. Comparative study of the two and four probe methods

In the first stage, a specimen without delamination is designed to check the validity of the four-probe method. The purpose of this study is to determine the validity range of the four probe method for the measurement of resistance in laminated composites.

The results of the surface and oblique resistances for various through-thickness conductivities are listed in Table 1 and plotted in Figures 6 and 7. In the Figures 6 and 7 the through-thickness conductivity is presented in the logarithmic scale. It is seen that the four-probe method is applicable for the measurement of resistance when the through-thickness conductivity does not exceed 2.5 $\Omega^{-1}mm^{-1}$. As the through-thickness conductivity decreases, the difference of obtained resistances by two- and four-probe methods significantly increases. For commonly used composite materials, the through-thickness conductivity is much smaller than the longitudinal conductivity. Thus, the four-probe method cannot be effectively used to measure the surface and oblique resistances for real composite materials. This result confirms the research of Todoroki [8] based on measurements of voltage in the through-thickness direction for plate type specimens.

Table 1. Comparison of electrical resistances between the two- and four-probe methods.

σ_t (Ω^{-1}mm^{-1})	Surface resistance R_{A1A5}, Ω		Oblique resistance R_{A1B5}, Ω	
	Two probe	Four probe	Two probe	Four probe
0.01	11.95	4.07	12.01	4.17
0.1	4.24	2.71	4.24	2.71
0.5	3.18	2.65	3.18	2.65
1	2.97	2.64	2.97	2.64
2.5	2.79	2.63	2.79	2.63
5	2.71	2.61	2.70	2.61
10	2.65	2.60	2.65	2.60
15	2.62	2.58	2.62	2.58

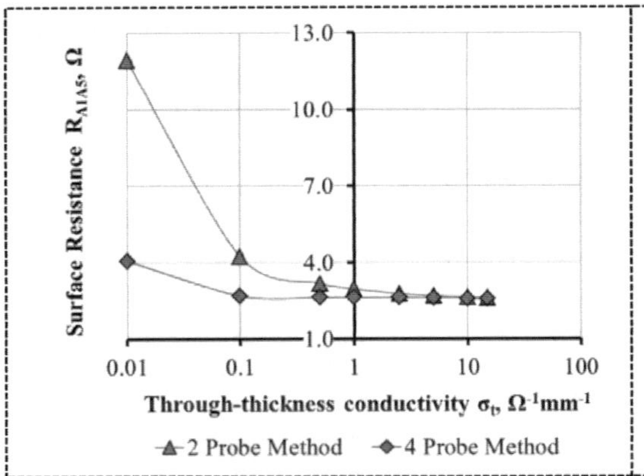

Figure 6. Comparison of the surface resistances between the two- and four-probe methods.

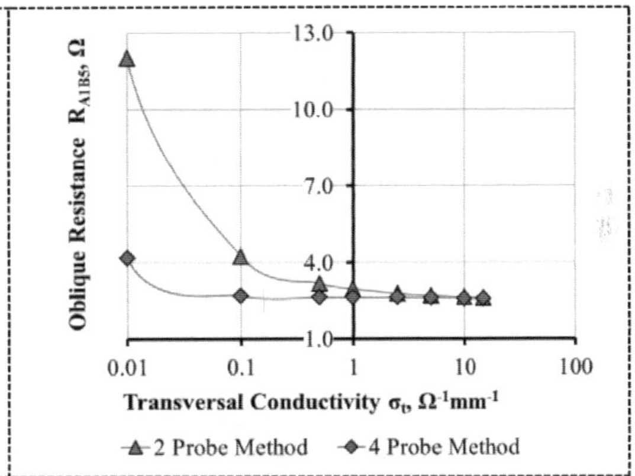

Figure 7. Comparison of the oblique resistances between the two- and four-probe methods.

4.2. The voltage and resistance percentage change due to delamination

In the next stage, the specimens with delamination are introduced and the similar computation procedure is used to obtain the electrical resistance at the segment with delamination by the two- and four-probe methods. The through-thickness conductivity σ_t of the composite specimens is varied from 0.01 Ω^{-1} mm^{-1} to 15 Ω^{-1} mm^{-1}. The length l of delamination is varied also starting with 20 mm up to 40 mm. The resistance percentage change is calculated using the ratio of the corresponding resistances for specimens with and without delamination as follow:

$$R_{\%} = \left(1 - \frac{R_0}{R_d}\right) \times 100\% \tag{2}$$

where R_d is the electrical resistance of a specimen with delamination and R_0 is the electrical resistance of a specimen without delamination.

Surface and oblique resistance percentage changes obtained by the two-probe method are given in Figures 8 – 11. One can see that the resistance percentage changes differ significantly depending on the current contacts used. For example, the resistance percentage change between electrodes A_0 and A_6 for both the surface and oblique resistance cases does not exceed 1.5%, while the resistance percentage change obtained between electrodes A_1 and A_5 gives a comparably larger value – 9%. This is explained by the fact the electrode A_1 lies right above the introduced delamination, and thus significantly affecting the electrical resistance in this segment of the composite panel.

IOP Conf. Series: Materials Science and Engineering **111** (2016) 012010 doi:10.1088/1757-899X/111/1/012010

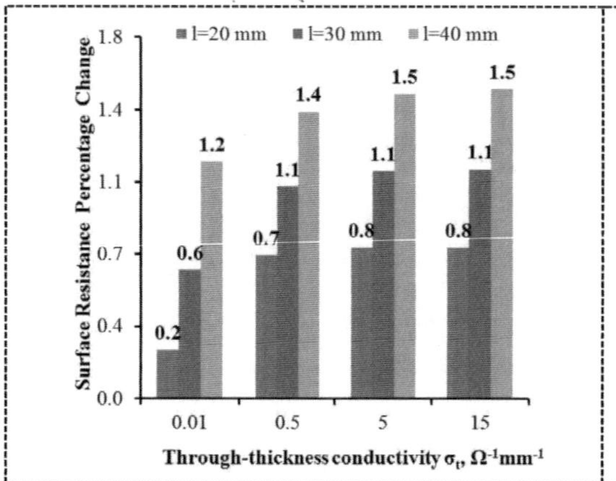

Figure 8. Surface resistance percentage changes in the two-probe method when location of current electrodes is A_0A_6.

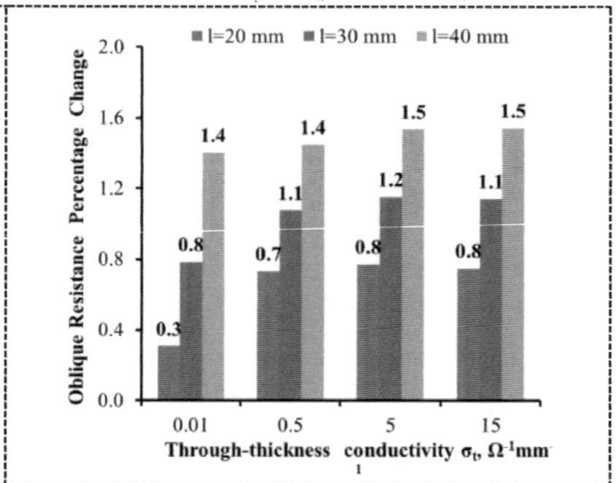

Figure 9. Oblique resistance percentage changes in the two-probe method when location of current electrodes is A_0B_6.

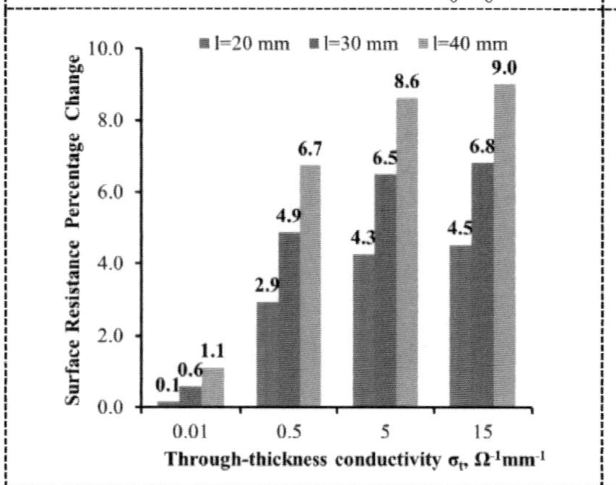

Figure 10. Surface resistance percentage changes in the two-probe method when location of current electrodes is A_1A_5.

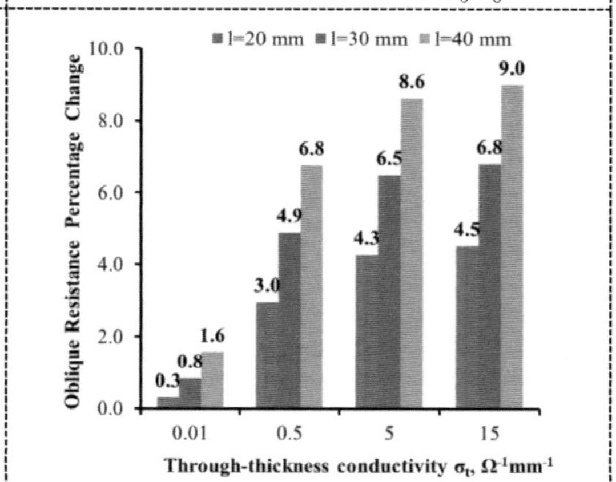

Figure 11. Oblique resistance percentage changes in the two-probe method when location of current electrodes is A_1B_5.

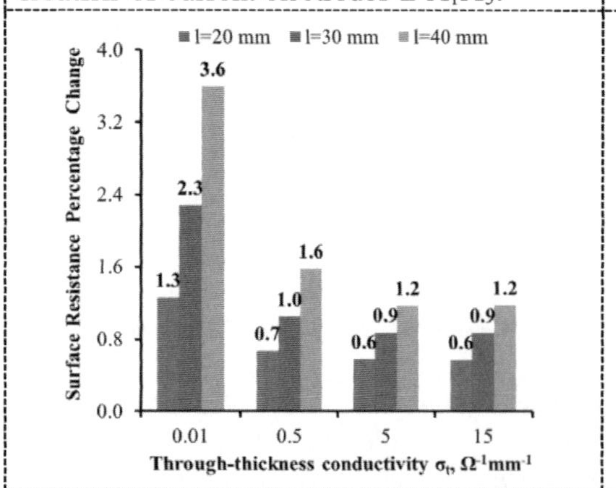

Figure 12. Surface resistance percentage changes in the four-probe method when location of current electrodes is A_0A_6 and location of voltage electrodes is A_1A_5.

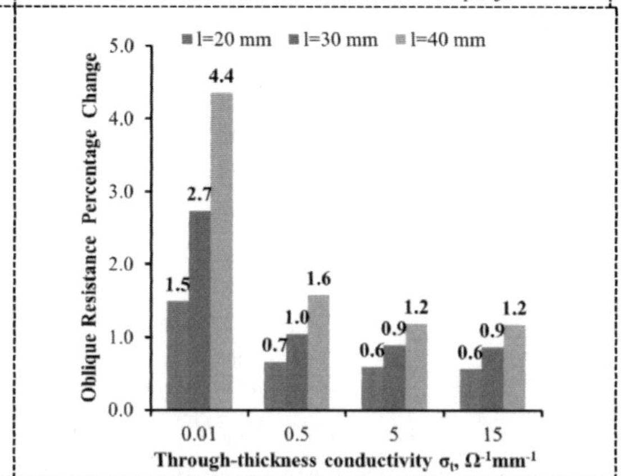

Figure 13. Oblique resistance percentage changes in the four-probe method when location of current electrodes is A_0B_6 and location of voltage electrodes is A_1B_5.

The corresponding resistance percentage changes are also obtained by using the four-probe method where the outside electrode pairs A_0A_6 or A_0B_6 are used as the current contacts and inside electrode pairs A_1A_5 or A_1B_5 as the voltage contacts (Figures 12-13). As it can be seen from these figures for the case when the through-thickness conductivity is the smallest one used in this study, the resistance percentage changes are the largest ones and they decrease as through-thickness conductivity increases. By comparing the two methods used in this study it must be noted that the two probe method works better for detection of delamination in case when material tends to be homogenous while the four-probe method is very effective when the through-thickness conductivity of a material is significantly smaller as it is for commonly used composite materials.

5. Conclusion

A 2-D numerical analysis of the electrical resistance for strip-type laminated composite specimens with and without delamination is carried out. The surface and oblique resistances are numerically calculated according to the two- and four-probe methods. The results obtained show that the four-probe method for the resistance measurements is valid only when the through thickness conductivity is comparable to the longitudinal conductivity. However, for the delamination detection in commonly used composite materials the four-probe method is more effective. The present study shows that the resistance percentage change is dependent on the location of electrode pairs used for the resistance measurement with the respect to the location of delamination. Thus, the resistance percentage changes could be used not only for the detection of the presence of delamination but also for the localization of delamination.

Acknowledgments

The present work was supported by the European Social Fund within the project 2013/0017/1DP/1.1.1.2.0/13/APIA/VIAA/06

References

[1] Liang D 1993 Fibre optic silicon impact sensor for application to smart skins. *Electron Lett* **29**(6) 529-530

[2] Haywood J, Coverley PT, Staszewski WI, Worden K 2005 An automatic impact monitor for a composite panel employing smart sensor technology *Smart Mater Struct* **14**(1) 265-271

[3] Shen L, Lib J, Liawb BM, Delaleb F, Chung JH 2007 Modeling and analysis of the electrical resistance measurement of carbon fiber polymer–matrix composites *Composites Science and Technology* **67**(11-12) 2513-2520

[4] Wang S, Wang D, Chung DDL 2006 Method of sensing impact damage in carbon fiber polymer-matrix composite by electrical resistance measurement *J Mater Sci* **41**(8) 2281-2290

[5] Wang S, Chung DDL, Chung J 2006 Self-sensing of damage in carbon fiber polymer-matrix composite cylinder by electrical resistance measurement *J Intell Mater Syst Struct* **17**(1) 57-62

[6] Todoroki A, Tanaka Y, Shimamura Y 2002 Measurement of orthotropic electric conductance of CFRP laminates and analysis of the effect on delamination monitoring with an electric resistance change method *Comput Sci Technol* **62** 619-628

[7] Todoroki A, Tanaka Y 2002 Delamination identification of cross-ply graphite/epoxy composite beams using electric resistance change method *Comput Sci Technol* **62** 629-639

[8] Todoroki A, Tanaka Y, Shimamura Y 2004 Multi-probe electric potential change for delamination monitoring of graphite/epoxy composite plates using normalized response surfaces *Comput Sci Technol* **64** 749-758

[9] Todoroki A, Tanaka Y, Shimamura Y 2005 Electrical resistance change method for monitoring delaminations of CFRP laminates: effect of spacing between electrodes *Comput Sci Technol* **65** 37-46

IOP Conf. Series: Materials Science and Engineering **111** (2016) 012011 doi:10.1088/1757-899X/111/1/012011

Nanocellulose reinforced polyurethane obtained from hydroxylated soybean oil

A Ivdre[1,2], V Mucci[3], P M Stefani[3], M I Aranguren[3] and U Cabulis[1]

[1]Polymer Laboratory, Latvian State Institute of Wood Chemistry, Riga, LV-1006, Latvia

[2]Riga Technical University, Faculty of Material Science and Applied Chemistry, Institute of Polymer Materials, P. Valdena Street 3, Riga, LV-1048, Latvia

[3]Institute of Research in Materials Science and Technology (INTEMA), National University of Mar del Plata - National Research Council (CONICET), Mar del Plata, Argentina

E-mail: aiga.ivdre@gmail.com

Abstract. In this research, cellulose nanocrystals were successfully obtained from microcrystalline cellulose and dispersed in a polyurethane matrix. Cellulose nanocrystals were characterized by dynamic light scattering. Polyurethane matrix was prepared from hydroxylated soybean oil, tertiary amine catalyst and polymeric diphenyl diisocyanate. Nanocellulose reinforced polyurethane composites were successfully prepared with different concentration and good dispersion of cellulose nanocrystals. Thermal properties and tensile strength of the resulting films were evaluated and dynamic mechanical analysis was carried out. The glass transition temperature of the nanocomposites increased slightly by increasing concentration of nanocellulose. The highest tensile strength was found for the nanocomposite with 1 wt% of cellulose nanocrystals.

1. Introduction

Polyurethane (PU) has become one of the most widely used plastics for various applications such as thermal insulation, construction, automotive parts and seating materials, in forms of rigid, semi-rigid and flexible foams as well as elastomers. Polyols for PU production are derived mostly from petrochemical products but can be replaced by renewable materials. The use of renewable resources in materials has attracted great interest in both industrial and academic fields because of their environmental aspect and sustainability [1, 2]. Polyols from such a natural oils as soybean [3-4], castor [5-6], palm oils [7-8], rapeseed [9-12] and sunflower oils [12-13] for obtaining PU materials has been widely investigated.

Also, different natural fillers could be used to improve characteristics of PUR, to reduce production costs and to enhance the renewable content in the product. The three major chemical components of biomass, cellulose, hemicellulose and lignin are utilized in diverse fields, such as biofuels, biomaterials, including conventional composites and novel nanocomposites, and other value-added chemicals [14]. Cellulose is a carbohydrate building block obtained from many prevalent resources to include wood, hemp, ramie, cotton, and bacteria [15]. It constitutes the most abundant renewable polymer resource available today. Most studies found in the literature about biocomposites use cellulose as simple “filler” which in some cases contributes to enhance the rigidity but mostly embrittles the polymer [16]. Through different processes cellulose can be separated into nano-scale forms [17-18]. The main process for the isolation of cellulose nanocrystals (CNC) from cellulose fibers is based on acid hydrolysis. When cellulose fibers are subjected to acid hydrolysis, the fibers yield defect-free, rod-like crystalline residues called nanowhiskers [17]. CNC is often being regarded as the next generation renewable reinforcement for the production of high performance biocomposites. CNC have been the subject of a wide array of research efforts as reinforcing agents in nanocomposites due to their low cost, availability, renewability, light weight, nanoscale dimension, and unique morphology [1, 19].

The present paper describes a method for obtaining well dispersed nanocomposites from hydroxylated soybean oil and different concentration of CNC. The mechanical and thermal properties of obtained nanocomposite films were investigated and compared.

2. Experimental

2.1. Materials

Polyurethane was prepared from the reaction of 4,4′-diphenylmethane diisocyanate prepolymer, pMDI (Rubinate 5005, Huntsman Polyurethanes, USA) with an equivalent weight of 131 g/eq and Agrol 3.6 (Biobased Technologies, USA) as the polyol component. This polyol was a hydroxylated soybean oil, OH value = 120.8 mg KOH/g (measured according to DIN 53240 norm). Dibutyl tin dilaurate (DBTDL) (Alkanos S.A. Co.) was used as the catalyst for the reaction. Commercial microcrystalline cellulose from Sigma Aldrich (USA) was acid hydrolized to prepare CNC. Other reagents used were: sulfuric acid, acetone (from Cicarelli, Argentina).

2.2. Synthesis and characterization of CNC

Acid hydrolysis of microcrystalline cellulose was carried out to obtain CNC. 15 g of microcrystalline cellulose and 52.5 g of distilled water were mixed in a flask by mechanical agitation. Afterwards 97.5 g of sulfuric acid (98 wt%) was added drop by drop while controlling that the temperature of the mixture did not exceed 30 °C. Once all the acid was added, the solution was heated to 44 °C. When temperature reached 44 °C, synthesis was continued for 2 hours more. The mixture was mechanically agitated during all the time of the synthesis. Then, the solution was cooled down in refrigerator for 24 h. The resulting solution has a pH of about 1-2. Dialysis against distilled water was then performed to remove any free acid molecules from the dispersion. Spectra/Por molecular porous membrane with molecular weight cut-off range of 12000 to 14000 Da was used for dialysis. Dialysis was continued until the pH of solution reached constant pH (pH ~6). After dialysis, the solution was filtered using a specific filter cloth to obtain nanocrystals by removing the last oversized cellulose particles.

Concentration of CNC was determined by gravimetric analysis. Size of particles were determined by dynamic light scattering.

2.3. Preparation of nanocomposites

The CNC solution was centrifuged 4 times using a cycle of 30 minutes at 12 000 rpm. At the end of each cycle the supernatant was removed and replaced with acetone. After centrifugation CNC dispersion in acetone was mixed in well dried Agrol polyol. Ultrasonic irradiation with IKA T 18 Basic Ultra-Turrax was used to disperse particles and to obtain homogenous dispersion. Afterwards acetone was evaporated by keeping mixture in vacuum for 2 days at 40 °C temperature.

Nanocomposites were prepared by mixing Agrol and CNC dispersion with DBTDL (0.1 wt%) and pMDI at NCO / OH molar ratio of 1.2. Mixture was poured in a closed mould which was pre-heated at 80 °C temperature. Afterwards the mould with sample was placed in a press at 0.5 MPa pressure and at 80 °C temperature. After 2 h the mould was put in an oven at 80 °C temperature for 16 h. All samples were allowed to stay at ambient conditions for a 24 h before cutting them into test samples. The concentrations of added CNC were 0, 0.5, 1.0 and 2.0 wt% with respect to the total weight of the sample.

2.4. Characterization of nanocomposites

Tests of dynamic mechanical analysis (DMA) (frequency of 1 Hz, amplitude of 0.1% and a ramp rate of 5°C/min.), tensile strength, and thermogravimetric analysis were carried out to characterize the PU nanocomposites. Tensile strength tests were performed with a load cell of 500 N at 1 mm/min and results were averaged out from five test specimen data.

3. Results and discussion

3.1. Size distribution of CNC

Results of dynamic light scattering showed that size of CNC is in range from 33 to 950 nm (Figure 1).

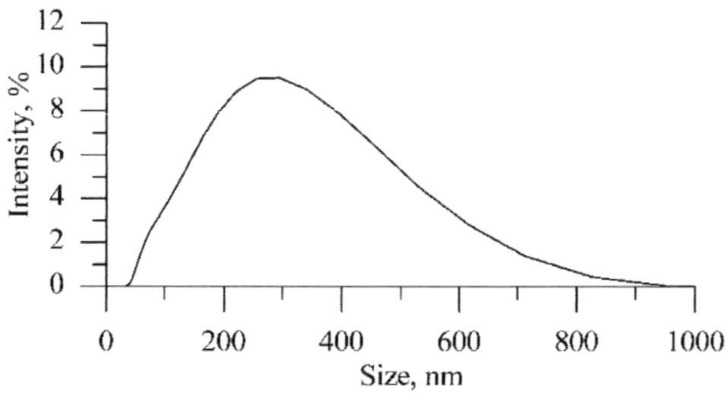

Figure 1. CNC particles size distribution

Size distribution showed peak at 295 nm, and the average size of CNC was ~200 nm which coincides with other publish results. Length of CNC can vary from 35 to 500 nm and width – from 3 to 48 [18, 20-21]. The size of CNC varies widely, depending on the source of the cellulosic material and the conditions (temperature, agitating, time etc.) under which the hydrolysis is performed [17].

3.2. Dynamic mechanical analysis and thermogravimetric analysis

Glass transition temperature (T_g) was measured from DMA results. Although there are several thermal techniques available to make T_g measurements, by far the most sensitive technique is DMA [22]. The maximum of the peak in the viscoelastic parameter tan δ was used to define the T_g. As shown in Table 1, the T_g of the nanocomposites increased slightly, reaching 35 °C for sample with 2 wt% of CNC.

Table 1. Glass transition temperature and initial decomposition temperature of nanocomposites.

CNC content, wt%	T_g, °C	T_{in}, °C
0	31	303.8
0.5	31	307.7
1.0	33	307.6
2.0	35	307.7

Increase of T_g indicates a favourable fiber–matrix interaction and restriction of molecular mobility because of the strong association between CNC and the soft segment of the PU matrix [1]. Results of thermogravimetry showed that thermal stability of nanocomposites is not significantly affected by incorporation of CNC (Table 1). The initial decomposition temperature (T_{in}) of PU composites (~308 °C) was a little bit higher than the neat sample without CNC (~304 °C).

3.3. Tensile strength and elastic modulus

Klemm et al. [23] found that the optimal enhancement of mechanical properties usually occurs at the point at which just enough reinforcing agent has been added to establish connectivity. For nanocellulose reinforced segmented PU films, *Auad et al.* obtained approximately 53% increase in tensile modulus by the addition of just 1.0 wt% CNC [24]. *Li et al.* investigated cellulose nano whiskers as a reinforcing filler in PU foams. Results showed that tensile modulus decreases by adding 0.5 wt% cellulose nano whiskers but increases by adding 1 wt% [3]. Also, *Floros et al.* investigations showed that tensile strength do not increase linearly by adding cellulose-based fibers to PU matrix. The best results were obtained with 0.5 wt% and 1.0 wt% CNC in the PU, but nanocomposite samples with CNC content beyond 1.0 wt% showed a negative effect on mechanical properties [25]. The mechanisms responsible for the improved mechanical performance are likely related to the nature of the nanoparticle and PU matrix.

Results of tensile strength and elastic modulus of our investigation are presented in Figure 2.

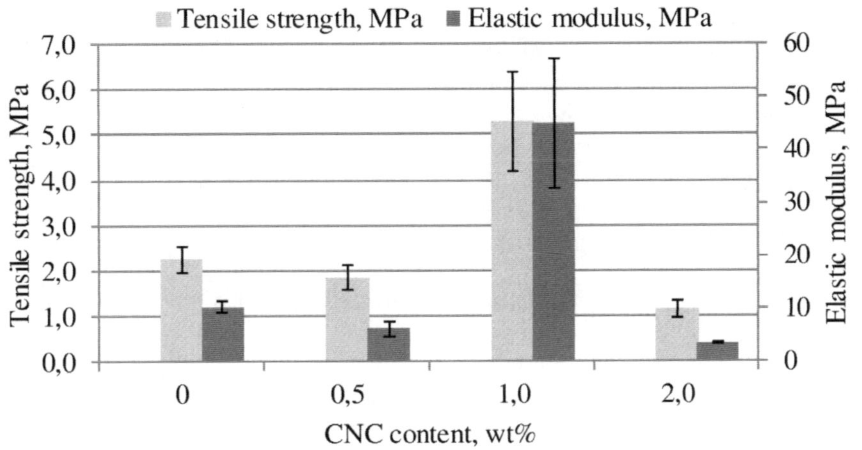

Figure 2. Tensile strength and elastic modulus of PU composites

In our study the tensile strength and elastic modulus increased more than 2 times and 4 times respectively by adding 1 wt% CNC in PU matrix reaching 5.3 MPa of tensile strength and 45 MPa of elastic modulus. However, samples with 0.5 wt% and 2.0 wt% CNC showed worse performance in mechanical properties than sample without CNC. Character of the change of tensile strength and elastic modulus coincides with other authors results described previously. Nanocrystals added in low concentration behave as point of failure in the composite, and probably interfere in the PU formation, since OH in the cellulose can also react with the isocyanate. At about 1 wt% of CNC the percolation of the system is reached, a network of touching crystals interpenetrates the PU matrix and the properties are much improved, although with a large dispersion of the results which is understandable because of the critical concentration (percolation threshold). Higher concentrations of the crystals result in agglomeration (concentrations above percolation) and thus the properties drop again, since the sample becomes heterogeneous and agglomerates are stress concentrators and point of failure for the material.

4. Conclusions

The objective of this study was to prepare PU composites with CNC and to examine how the concentration of CNC affects the final properties of PU composites. CNC were successfully obtained from microcrystalline cellulose by acid hydrolysis and dispersed in a PU matrix based on hydroxylated soybean oil. Nanocellulose reinforced PU composites were successfully prepared with different concentration (0 – 2 wt%) and good dispersion of CNC.

T_g of the nanocomposites increased slightly, reaching 35 °C for sample with 2 wt% of CNC. Results of thermogravimetric analysis showed that thermal stability of nanocomposites is slightly increased by incorporation of CNC. Results of tensile strength and elastic modulus showed that 1 wt% of CNC is the optimal concentration to improve mechanical properties of PU. The tensile strength increased more than 2 times comparing with neat sample without CNC.

Overall, results indicated that CNC can improve final properties of PU composites based on hydroxylated soybean oil. Further investigations are required to explore CNC effect onto other properties of PU composites.

5. Acknowledgments
The financial support of BIOPURFIL project (agreement number: PIRSES-GA-2012-318996) by the EC through the FP7 is gratefully acknowledged.

6. References

[1] Park S H, Oh K W and Kim S H 2013 *Compos. Sci. Technol.* **86** 82
[2] Ionescu M 2008 *Chemistry and Technology of Polyols for Polyurethane* Shawbury, Shrewsbury, Shropshire, UK, Smithers Rapra Press. 602
[3] Ji D, Fang Z, He W, Luo Z, Jiang X, Wang T and Guo K 2015 *Ind. Crops. Prod.* **74** 76
[4] Lee A and Deng Y 2015 *Eur. Polym. J.* **63** 67
[5] Zhang L, Zhang M, Hu L and Zhou Y 2014 *Ind. Crops. Prod.* **52** 380
[6] Fu C, Yang Z, Zheng Z and Shen L 2014 *Prog. Org. Coat.* **77** 1241
[7] Tanaka R, Hirose S and Hatakeyama H 2008 *Bioresource Techon.* **99** 3810
[8] Pauzi N P N, Majid R A, Dzulkifli M H and Yahya M Y 2014 *Compos. Part B-Eng.* **67** 521
[9] Stirna U, Fridrihsone A, Lazdiņa B, Misāne M and Vilsone D 2013 *J. Polym. Environ.* **21** 952
[10] Kurańska M, Prociak M, Kirpluks M, Cabulis U. 2015 *Ind. Crops. Prod.* **74** 849
[11] Paberza A, Fridrihsone-Girone A, Abolins A, Cabulis U. 2015 *Polimery* **60** 572
[12] Stirna U, Cabulis U, Beverte I 2008 *J. Cell. Plast.* **44** 139
[13] Das B, Konwar U, Mandal M and Karak N 2013 *Ind. Crops. Prod.* **44** 396
[14] Li Y and Ragauskas A 2011 *Advances in Diverse Industrial Applications of Nanocomposites* **2** 17
[15] Moon R J, Martini A, Nairn J, Simonsen J and Youngblodd J 2011 *Chem. Soc. Rev.* **7** 3941
[16] Bras J, Hassan M L, Buzesse C, Hassan E, El-Wakil N and Dufresne A 2010 *Ind. Crops. Prod.* **32** 627
[17] Habibi Y, Lucia L A, Rojas O J 2010 *Chem. Rew.* **110** 3479
[18] Elazzouzi-Hafraoui S, Nishiyama Y, Putaux J L, Heux L, Dubreuil F and Rochas C 2008 *Biomacromolecules* **9** 57
[19] Lee K, Aitomäki Y, Berglund L A, Oksman K and Bismarck A 2014 *Compos. Sci.Technol.* **105** 15
[20] Pranger L and Tannenbaum R 2008 *Macromolecules* **41** 8682
[21] Capadona J R, Shanmuganathan K, Trittschuh S, Seidel S, Rowan S J and Weder C 2009 *Biomacromolecules* **10** 712
[22] Thermal analysis and reology http://www.tainstruments.com/pdf/literature/TS64.pdf (access date 14.09.2015)
[23] Klemm D, Kramer F, Moritz S, Lindstrom T, Ankerfors M, Gray D, Dorris A 2011 *Angewandte Chemie-International Edition* **50** 5438
[24] Auad M L, Contos V S, Nutt S, Aranguren M I, Marcovich N E 2008 *Polym. Int.* **57** 651
[25] Floros M, Hojabri L, Abraham E, Jose J, Thomas S, Pothan L, Leao A L and Narine S 2012 *Polym. Degr. Stabil.* **97** 1970

Properties of Composites Based on Linear Low Density Polyethylene (LLDPE) and by Sol-gel Method Modified Fibres

Z Zelca[1], S Kukle[2], J Kajaks[3], I Rucevska[4], S Vihodceva[5]

[1,2,4,5]Riga Technical University, Faculty of Material Science and Applied Chemistry, Institute of Design Technologies, Kipsalas Street 6, Riga, LV-1048, Latvia
[3]Riga Technical University, Faculty of Material Science and Applied Chemistry, Institute of Polymer Materials, P. Valdena Street 3, Riga, LV-1048, Latvia

E-mail: zane.zelca@rtu.lv

Abstract. The influence of sol-gel modified fibres content on the melt flow index (MFI) and water absorption of a linear low density polyethylene was investigated. The fibres were modified with silica xerogel prepared by the sol-gel polymerization of tetraethylorthosilicate (TEOS) under hydrolytic conditions using acid catalysis. Alkali pre-treatment was used for some variants and some composite samples were prepared by adding maleic anhydride grafted polyethylene (MA-g-PE). It was found out that melt flow index increases using sol-gel method for hemp fibres modification with fibres content in composite 40 and 50 wt%. Composite with fibres modified by sol-gel method content 40 wt% water absorption were evaluated in a range from 8.5 to 16.32%, but for composites with 50 wt% - 12.12%.

1. Introduction

Optimal content (30 wt%) of fibres in LLDPE matrix and unmodified hemp fibres reported Solizenko et.al. These composites maintain sufficient fluidity, and they can be processed by traditional polymer processing methods [1]. The fibres content seems relatively small to meet growing necessity of renewable recourses integration in materials.

For composites make more environmentally friendly, it is necessary to increase the fibres content and one of the options is chemical treatments like mercerization, acetylation and others. It is important to impart hydrophobicity to the hemp fibres by suitable chemical treatment to develop composites with better mechanical properties and decrease water absorption.

Vihodceva and co-authors reported about sol-gel technology using for the pure cotton textile treatment for imparting of such properties as stable water-repellency, antimicrobial, and UV protection shows a high wear resistant [2].

For composites production with high hemp fibres content is essential such property as a weak water absorption in this research was evaluated the possibility of hemp fibres treatment by mentioned above sol-gel technology with proved water-repellent imparting properties. Hemp fibres, the same as cotton fabrics need to adjust the highest possible heat treatment, which does not damage the fibres and provides the necessary properties of the coating.

In the article the hemp fibres treatment by reported sol synthesis method before composite formation is described to reduce water absorption and measure the sol treatment effect and other treatment combination influence on composite MFI.

2. Materials and Methods

The hemp fibres used in this study were obtained from hemp stems harvested from a trial plot at Agriculture Science Centre of Latgale in Vilani district, Latvia. Hemp fibres of variety 'Bialobrzeskie' in growing process were cultivated with active nitrogen fertilizer (N30). The harvested hemp stems were left for dew retting on the field 4 weeks.

Content of hemp fibres (length up to 1 mm) in composites was 40 and 50 wt% with pre-processing NaOH (SIA"Enola", ES) 2% solution (4h +23C^0) and sol-gel method. Materials used for sol synthesis- $C_8H_{20}O_4Si$ (Alfa Aesar, German), C_2H_5OH and HF (ES (SIA"Enola", ES), deionized water, with or without adding $Zn(CH_3COO)_2{\bullet}2H_2O$ (Scharlau, Spain) used as the modifier of the sol-gel system and method reported by Vihodceva et.al.

LLDPE grade LL 6201 was used as matrix, MFI = 50g/10min, $T_{melt.}$ = 123^{0}C.

MA-g-PE (Licocene fine grain TP 4351) was used as a coupling agent 5 wt%.

Before mixing, the fibres were dried in an air chamber at +105^{0}C for 24h.

Composites were prepared by mixing of the components on two rolls mill (T=150^{0}C 10 min), then cooled and pressed in 1mm thick sheets. Water exposure experiments (standard ASTM D 570-98) were carried out at room temperature (+23C^0). Fluidity was estimated by melt flow index method (MFI) according to the standard ASTM D 1238-90b (T=190^0 C, P=2.16 kg).

Using Atomic force microscopy composite cross-section surface was investigated. The reference voltage was set to 2.6+/-0.2V, a level that provided clear images. Test operated in Tapping mode, image size 25x25 μm and 20x20 μm with Scan Rate 0,1 Hz. Silicone probe (OTESPA-R3) nominal tip radius, Spring constants and free resonance frequencies were respectively 3.75 μm, 26 N/m and 300kHz.

10 composite samples were prepared with different hemp fibres treatment (Tab 1.).

Table 1. Composite samples

Sample	Hemp fibres processing method	Fibres content, wt%
2a	Sol-gel	40
2b	Sol-gel and MA-g-PE 5 wt%	40
3a	Sol-gel with $Zn(CH_3COO)_2{\cdot}2H_2O$5 wt%	40
3b	Sol-gel with $Zn(CH_3COO)_2{\cdot}2H_2O$ 5 wt% and MA-g-PE 5 wt%	40
4a	2% NaOH and Sol-gel	40
4b	2% NaOH and Sol-gel and MA-g-PE 5 wt%	40
5a	2% NaOH and Sol-gel with 5 wt% $Zn(CH_3COO)_2{\cdot}2H_2O$	40
5b	2% NaOH and Sol-gel with 5 wt% $Zn(CH_3COO)_2{\cdot}2H_2O$ and MA-g-PE 5wt%	40
6	Without pre-treatment	50
7a	Sol-gel	50

3. Results and Discussion

Melt-flow indexes of variants fall in a range from very low (0.1 for samples with un-processed fibres) till very high (24.41 for samples with silica sol pre-treated fibres).

Untreated hemp fibres composite with 40 wt% practically lost fluidity [3]. Previous experiments 10% solution of acetic anhydride hemp fibres treatement 40 wt% composite MFI increased to 8.47g/10min but 50 wt% composite to 1.81g/10min [4]. Sol-gel method hemp fibres modification shows magnificent fluidity increase, more than 20 times (sample 2a, 2b, 4b) composite with hemp fibres 40 wt% and increasing fibres content to 50 wt% no significant difference (sample 7a), Fig.1.

For better adhesion between by sol-gel method obtained coating and hemp fibres surface some samples used alkali pre-treatment. Previous experiments showed fluidity decrease for about 2 times using 2% NaOH fibres treatment [4]. Combination of alkali pre-treatment with Sol-gel treatment shows 22% fluidity decrease (2a and 4a) compared to only sol-gel treatment. MA-g-PE additive for sample with alkali pre-treatment showed MFI increase 21% (4a to 4b) and sample with $Zn(CH_3COO)_2{\cdot}2H_2O$ modifier of the sol-gel system MFI increase 52% (5a and 5b).

Samples with Sol-gel modified fibres with $Zn(CH_3COO)_2{\cdot}2H_2O$ modifier of the sol-gel system (3a and 3b) showed lower results but processing of the composites is possible by traditional methods.

Two rolls mill composite processing can pull-of particles from sol layer on hemp fibres surface and could be incorporated into polyethylene matrix. Therefore silica polymer nano-composite nanoparticles can modify matrix and affect some polyethylene properties in the next processing stages. Silica polymer nano-composites agglomerates presence can significantly reduce the silica loading, resulting in reduced thermochemical properties.

Dorigato et. al. reported about linear low density polyethylene/silica nano-composites crosslinking reaction induced by thermal processing (200^{0}C) caused a remarkable increase of the melt viscosity, as revealed by the melt flow index values of both neat matrix and nano-composites [5].

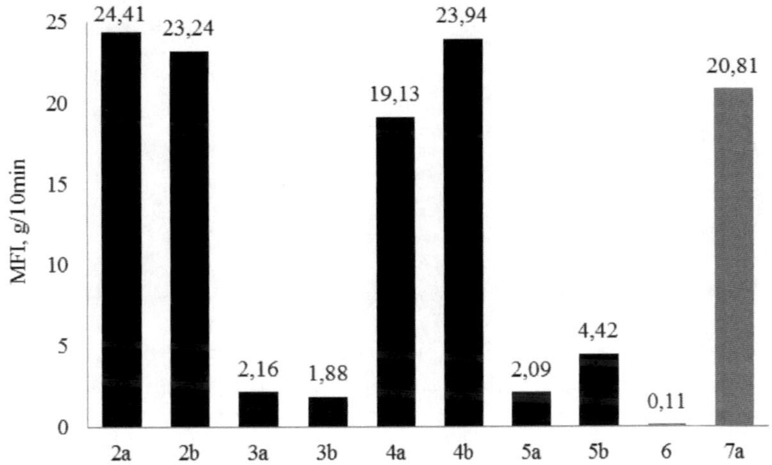

Figure 1. Melt flow index

At lower temperatures that are used for the hemp fibres thermal treatment (about 120^{0}C) starts partial decomposition of zinc acetate dehydrate and basic zinc compound formation [6]. In the formation of the zinc acetate ($C_4H_6O_4Zn$), silica compounds: $H_2Si_3O_7(H_2O)$ and $H_2Si_{14}O_{29}{\bullet}5.4H_2O$, and probably from insoluble basic zinc compounds or zinc oxide in amorphous phase can react polymer chains in composite preparing process. Crystalline phase of the ZnO appears starting from the temperature 300°C, in the range of the temperature from 300 to 500°C SiO_2 is also obtained; from the 400°C the formation of the willemite - $Zn_2(SiO_4)$ begins [6], therefore no crystalline phase Zn and Si particles into the matrix possible. Additional studies are required to determine how affects xserogel amorphous phase to the matrix viscosity.

Water sorption after 438h (Fig. 2. and Fig. 3) varies in range 8.5 (40 wt% sample 4b) to 22.16 (50wt% sample 6 with un-processed fibres) and increases with the enhance of the fibres content from 40 to 50 wt%.

Composite water absorption with untreated hemp (40 wt%) fibres varies in range 10.8 to 12.10% (240h) [1].

The highest water resistance showed in the sample 4a was fibres pre-treated with NaOH and then modified fibres surface with sol-gel method. Alkali pre-treatment clean fibres and provides sustainable xerogel coating that is sufficiently stable after insertion in polythene.

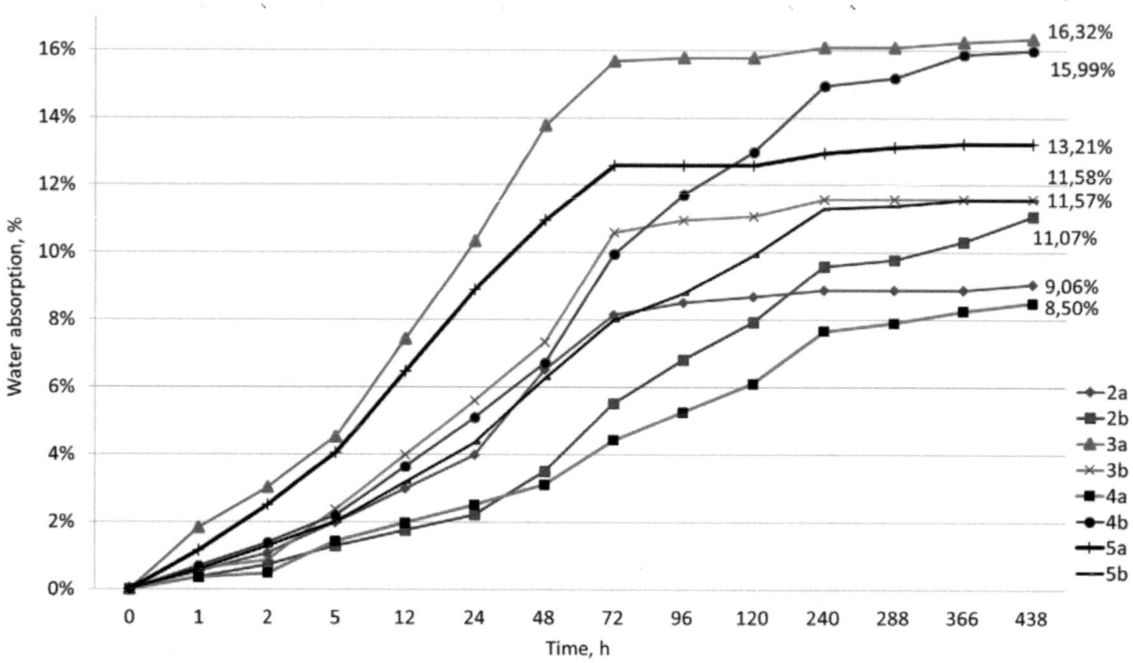

Figure 2. Hemp fibres (40wt%)/LLDPE composite water absorption

But sample 4b with 5 wt% MA-g-PE additive showed one of the lowest water resistance, this indicates that the xerogel coating form chemical bonds with the MA-g-PE as coupling agent and lost the resilience of the fibres. All samples except 4b show higher water resistance with MA-g-PE than without additive.

After 438 hours water exposure for samples of the 3a, 3b, 5a, 5b composite variants equilibrium moisture content established.

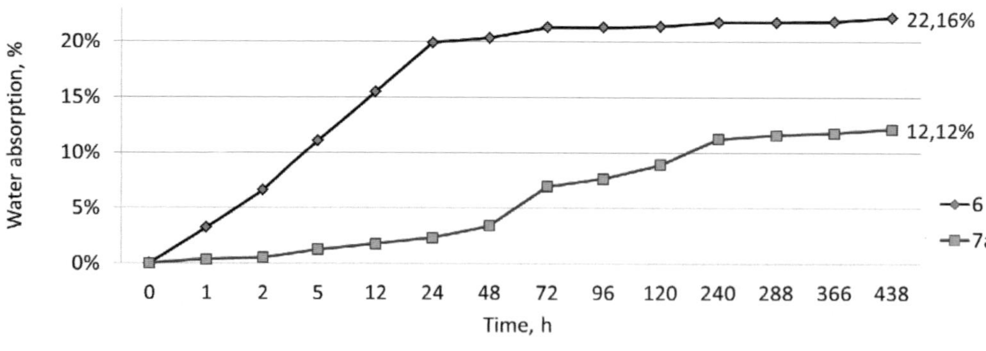

Figure 3. Hemp fibres (50wt%)/LLDPE composite water absorption

Composite water absorption (7a) with silica xerogel treated hemp fibres 50 wt% is 12.12% and it is in the same range as untreated hemp fibres 40 wt% composite water absorption. Hemp fibres (50wt%)/LLDPE composite equilibrium moisture content not established, water uptake is still in process.

In cross section of 50 wt% hemp fibres and LLDPE composite (Fig.4.) are seen compact packed with evenly spread compounds, cannot see cracks and air inclusions. But Fig.5. hemp fibres detached from matrix and not evenly dispersed.

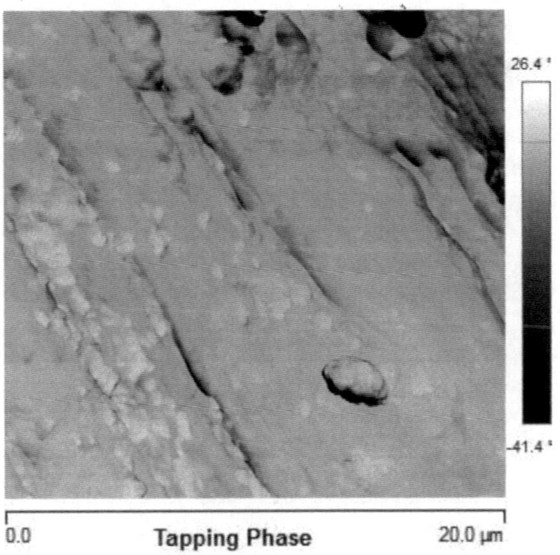

Figure 4. Cross section of 50 wt% hemp fibres/ LLDPE composite (7a)

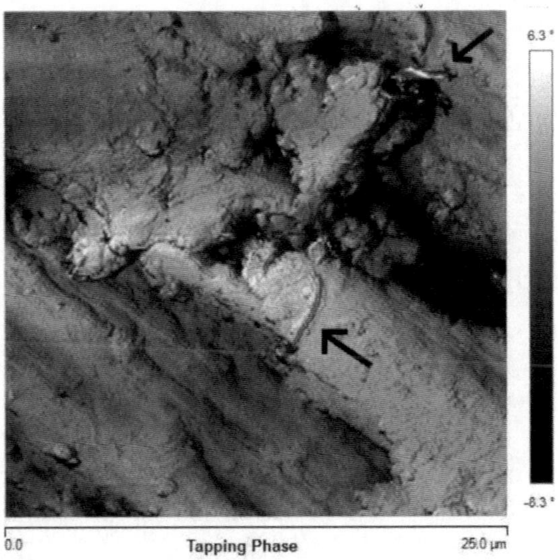

Figure 5. Cross section of 50 wt% hemp fibres/ LLDPE composite (6)

4. Conclusions

The melt flow index increases using sol-gel method for hemp fibres modification with fibres content in composite 40 and 50 wt%. $Zn(CH_3COO)_2{\cdot}2H_2O$ the modifier of the sol-gel system decrease MFI results but do not give significant improvements in water absorption. MA-g-PE additive increase water resistance but decrease MFI except combination with alkali pre-treatment.

Using sol-gel method for hemp fibres modification in LLDPE composite matrix can increase filler content 50 wt% and more maintaining fluidity and improve water resistance.

The Sol-gel method is suitable to modify hemp fibres hydrophobic surfaces and decrease water absorption and increase MFI.

Future studies need to explore the xerogel coating promising effects on matrix and sol-gel method effect on composite mechanical properties.

References

[1] Solizenko R, Kajaks J, Nestore O, Kukle S 2013 The Exploitation Properties of Hemp Fibres Containing Linear Low Density Polyethylene Composites. *Mat. Sci. and App. Chem.* Vol.27, 17-21 pp ISSN 1407-7353 print / 2255-8713 online

[2] Vihodceva S, Kukle S 2012 Cotton Fabric Surface Modification by Sol-Gel Deposition of ZnO Thin Films, IOP Conf.*Ser.:Mater. Sci. Eng.* **38** 012022 doi:10.1088/1757-899X/38/1/012022

[3] Nestore O, Kajaks J, Vancovicha I, Reihmane S 2013 Physical and mechanical properties of composites based on a linear low-density polyethylene (LLDPE) and natural fibres waste, *Mech. of Comp. Mat.* Vol. 48, Issue 6, pp 619-628

[4] Kajaks J, Zelca Z, Kukle S 2015 The Effect of Modification Methods on Performance Characteristics of Composites based on Linear Low Density Polyethylene (LLDPE) and Natural - Hemp Fibres, *Mech. of Comp. Mat* .Vol. 51, Issue 5, pp 955-964 ISSN 0191-5665 print/ e-ISSN 1573-8922 online

[5] Dorigado A, Pegoretti A 2013 (Re) processing effects on linear low-density polyethylene/silica nanocomposites *J. Polym. Res.* ISSN 1022-9760 print/1572-8935 online doi:10.1007/s10965-013-0092-x

[6] Vihodceva S, Kukle S and Bitenieks J 2015 Durable hydrophobic sol-gel finishing for textiles IOP Conf. Ser.: *Mater. Sci. Eng.* **77** 012023 doi:10.1088/1757-899X/77/1/012023

IOP Conf. Series: Materials Science and Engineering **111** (2016) 012013 doi:10.1088/1757-899X/111/1/012013

CHARACTERIZATION OF BARK RICH-IN-TANNINS EXTRACTS FROM DECIDUOUS TREES WITH EMPHASIS ON THEIR ANTIOXIDANT ACTIVITY

Janceva S[1], Dizhbite T[1], Vevere L[1], Arshanitsa A[1], Telysheva G[1], Dzenis M[2]

[1] Latvian State Institute of Wood Chemistry, Dzerbenes Street 27, Riga, LV-1006, Latvia

[2] Riga Technical University, Faculty of Material Science and Applied Chemistry, Institute of Polymer Materials, P. Valdena Street 3, Riga, LV-1048, Latvia

Abstract. The barks of deciduous trees spread over a wide area of Northern Europe, were sequentially extracted using solvents of increasing polarity. The highest yields of hydrophilic extractives were found for ash tree and goat willow (23.6% and 23.8%, respectively). The highest total contents of polyphenols were found for black alder and goat willow bark (0.18 GAE·g^{-1} and 0.12 GAE·g^{-1}, respectively). Hydrophilic extract from the both grey and black alder barks contained high amount (up to 7 % on bark dry mass) of condensed tannins (CT) or oligomeric proanthocyanidins (OPC). In tests with free radicals (ABTS$^{\bullet+}$, DPPH$^{\bullet}$), the high radical scavenging capacities of the hydrophilic extracts enriched with OPC were demonstrated. The antioxidant efficiency of the hydrophilic extracts and purified OPC from alder barks was tested by their influence on thermo-oxidative destruction of model polyurethane (PU) films. The hydrophilic extracts enriched with tannins have good potential as a technical antioxidant for polyurethanes, with the most prominent activity for the hydrophilic extract from black alder and goat willow bark.

1. Introduction

Up till now wood bark from logging and wood mechanical processing is used as cheap fuels. Unique compositional characteristics of bark biomass and possibility of numerous value added products (individual compounds and mixtures of synergetic activity) obtaining promote researches aimed at installation of bark utilization in integrated biorefinery schemes of wood processing. The barks of deciduous tree species are known to contain bioactive chemicals, including the OPC, which belong to the group of polymeric polyphenols. Accumulation of polyphenols in bark is the results of plants evolutions as their response on biotic and abiotic stresses. The high concentration of OPC in trees bark makes it a prospective raw material not only for medicine and veterinary but also for industrial applications. The bark of deciduous trees species growing in the Northern Eirope, in particular, in Latvia, is scarcely explored as a source for obtaining of valuable extractives products [1]. Polymeric polyphenols pool of bark is considered to be beneficial for various technical applications, such as adhesive and binder, components of bio-composites, biocides for agricultural needs etc. Currently the commercialized products based on rich-in-tannins extracts from bark are manufactured on the basis of tree species growing in Southern countries. CT quebracho and mimosa are produced commercially from quebracho tree (*Schinopsis lorentzii*) and black wattle (*Acacia mearnsii*) barks and are often used as a raw material for the industrial purposes. The antioxidant properties of OPC make their application very attractive. Demand for natural antioxidants is increasing, primarily due to their higher safety in comparison with synthetic compounds. Bark extracts are a natural resource that could replace synthetic antioxidants as butylatedhydroxyanisole (BHA) and butylatedhydroxytoluene (BHT), which are suspected of

being responsible for liver damage and carcinogenesis in organisms of laboratory animals [2]. As an antioxidant, PAC has two advantages: natural origin (much lower toxicity and bio-degrability) and oligomeric structure. The interest in high-molecular weight antioxidants is connected with the possibility to condition the long-term stability to polymeric composite materials [3]. At present, oligomeric and polymeric antioxidant are topics of great interest for researchers in many fields, such as pharmaceuticals, cosmetics, food and polymer composites production.
The aim of the present work was screening of some widely spread in Latvia and Northern Europe deciduous tree species: grey alder (*Alnus incana*), black alder (*Alnus glutinosa*), ash tree (*Fraxinus excelsior*), and goat willow (*Salix caprea*) as potential sources for obtaining of hydrophilic extracts enriched with proanthocyanidins and their testing as a polymeric antioxidant for polyurethane polymer compositions.

2. Materials and methods

Hydrophilic extracts enriched with OPC were isolated from bark using sequentially extraction with Accelerated Solvent Extractor (ASE 350, Dionex) at 90° C (1500 psi) for 20 min using solvents of increasing polarity: hexane, ethyl acetate and finally, aqueous ethanol (1:1, v/v). The total amount of extractives was determined gravimetrically. The yields of the fractions are shown as % of dry bark. Total polyphenols (TP) content was determined using Folin-Ciocalteu procedure [4]. OPC content was measured by buthanol-HCl method [5] using procyanidin dimer B2 as a reference compound. Py-GC/MS/FID analysis was used for characterization of extracts on the presence of carbohydrates. OPC were purified from non-tannin phenolics using Sephadex LH-20. TOF-MS (QSTAR Elite System Hybrid Quadrupole-TOF-MS) were used for characterization of OPC. The extracts enriched with OPC were tested for their antioxidant activities using $DPPH^{\bullet}$ and $ABTS^{\bullet+}$ radical scavenging assay. Free radical scavenging activity was expressed as IC_{50} (the concentration of antioxidant, $mg{\cdot}L^{-1}$, required for a 50% inhibition of the free radical). The lower is the IC_{50} value, the higher is the radical scavenging activity of the compounds. Trolox (a water-soluble derivative of vitamin E) was tested as a reference antioxidant. Effect of extracts and purified OPC on the thermo-oxidative destruction of PU elastomer (control and containing 2.0% extract) was studied in the temperature range of 20 to 200°C by TGA and DTA (thermogravimetry and differential thermal analysis) in an air atmosphere (flow rate of 50 mL min^{-1}) using a Perkin Elmer, STA 600 TGA/DTA system at a heating rate of 10°C min^{-1}. The commercial technical antioxidant Irganox was used in this trial as the reference antioxidant. The model PU elastomer films were obtained as described in Arshanitsa *et al.* [3].

3. Results and discussion

The highest yield of hydrophilic extractives were found for ash tree and goat willow (23.6 % and 23.8 %, respectively). The highest total contents of polyphenols (TP) were found for black alder and goat willow bark (0.18 $GAE{\cdot}g^{-1}$ and 0.12 $GAE{\cdot}g^{-1}$ of bark, respectively). Based on the Py-GC/MS/FID results, extracts from barks of all tree species contained polyphenols as the predominant component, although a certain amount of carbohydrates (15 – 28 % on extracts dry mass) are also present. Hydrophilic extracts from the both alder species contained high amount (up to 7% on bark dry mass) of OPC, whereas OPC content of extract from ash tree was negligible. The TOF–MS spectrum of proanthocyanidins from bark of alder species showed that the discharged compounds from deciduous trees barks are B-type proanthocyanidines pentamers with catechins or epicatechins (288 Da) degree of polymerization being 5 (Figure 1).

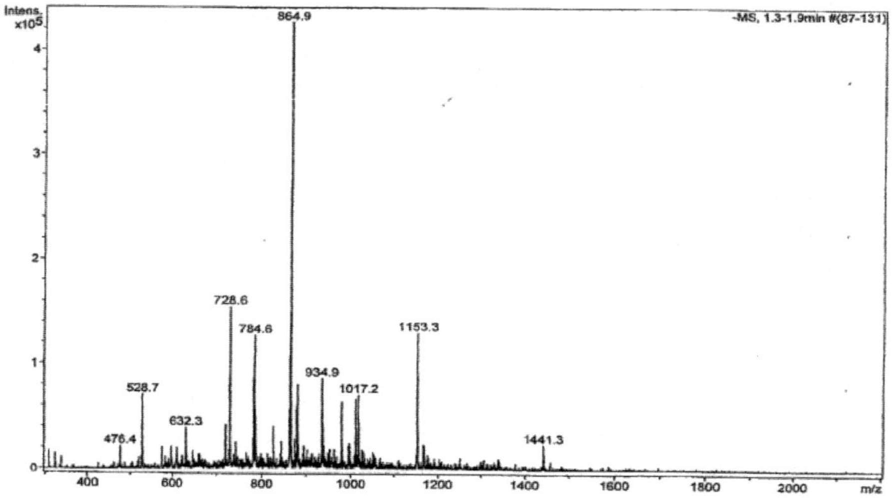

Figure 1. TOF –MS spectrum of OPC from grey alder bark

The assessment of antioxidant activity of deciduous tree barks extracts and OPC opens wider perspectives for rationalization of natural resources usage. The results of the tests on radical scavenging activity are presented in the Table 1. The comparison of the results of two tests (DPPH$^{\bullet}$ and ABTS$^{\bullet+}$) for hydrophilic extracts revealed the highest level of activity for the extract from bark of goat willow (IC_{50} of 1.8 mg·L^{-1}). Among purified OPC, the highest antioxidative activity, which was higher than that for Trolox, showed OPC from the bark of grey alder. The data obtained (Table 1) on carbohydrates content in extracts (Py-GC/MC/FID) showed that with an increase of carbohydrate content the antioxidant activity reduced.

Table 1

Antioxidant activity of rich-in-tannins extracts and purified OPC from barks of deciduous trees

Sample	TP content, GAE·g^{-1} *(Folin-Ciocalteu method)*	OPC content % on o.d. extract *(buthanol-HCl method)*	Carbohydrates content % on o.d. extract *(Py-GC/MC/FID)*	IC_{50}, mg·L^{-1}	
				ABTS$^{+\bullet}$	DPPH•
Goat willow	0.7	17.7	15.1	1.8	3.8
Black alder (BA)	0.7	25.9	24.3	3.0	6.1
Grey alder (GA)	0.6	25.2	28.3	3.8	7.5
Ash tree	0.2	0	n.d	8.7	33.6
OPC from GA	-	>90.0	<0.01	1.4	3.0
OPC from BA	-	>90.0	<0.01	3.3	6.9
Trolox	-	-	-	4.0	4.7

In order to evaluate an antioxidant activity of bark products under study they were introduced in compositions of model PU films in the amount close to the usually used in practice for technical antioxidants applications. The TGA data (Table 2) clearly showed the antioxidant effect of extracts and OPC revealed in increase in starting (T_{start}) and maximal rate (T_{max}) temperatures of PU thermo-oxidative destruction. The DTA data also confirmed the increase in thermo-oxidative stability of model PU films: the exothermal maximum connected with PU thermooxidation accompanied by volatiles release shifted towards the high-temperature region by 15-30°C.
The TGA data clearly revealed that the antioxidant effects of the hydrophilic extracts from grey alder and goat willow barks were on the same level as those for purified OPC and higher than that for commercial reference antioxidant Irganox used usually for PU systems.

Table 2
Thermal properties of rich-in-tannins extracts and purified OPC from bark of deciduous trees

Sample	T_{start}, °C	T_{max}, °C
PU without antioxidant	285	323
PU with 2 % GA extract	310	342
PU with 2 % BA extract	300	342
PU with 2 % GW extract	316	350
PU with 2% OPC from GA	319	341
PU with 2% OPC from BA	310	342
PU ar 2% Irganox	293	323

The research results have shown that the rich-in-tannins extracts from alder and goat willow barks are promising agents for stabilization of PU systems that enhance the high limit of PU exploitation temperatures.

Acknowledgement: The authors greatly appreciate the collegial help of M. Kuhtinskoya in characterization of OPC by TOF-MS analysis. Financial support from the Latvian budget, Government Research Program ResProd and LZP grant 564/2012 is gratefully acknowledged.

Reference

[1] Telysheva G, Dizbite T, Bikovens O, Ponomarenko J, Janceva S and Krasilnikova J 2011 Structure and antioxidant activity of diarylheptanoids extracted from grey alde (Alnus incana) and potential of biorefinery-based bark processing of European trees *Holzforshung* **65 (4)** 623-629
[2] Thompson D and Moldeus P 1988 Cytotoxicity of butylated hydroxyanisole and butylated hydroxytoluene in isolated rat hepatocytes *Biochem. Pharmacol.* **37 (11)** 2201-2207
[3] Arshanitsa A, Ponomarenko J, Dizhbite T, Andersone A et al 2013 Fractionation of technical lignins as a tool for improvement of their antioxidant properties *J. Anal. Appl. Pyrol.* **103** (SI) 78-85
[4] Waterhouse AL 2000 Determination of total phenolics *Food science* **2** 567
[5] Schofield P., Mbugua D M, Pell A N 2011 Analysis of condensed tannins: a review *Animalfeed sciece and tehnology* **91** 21-40.

IOP Conf. Series: Materials Science and Engineering **111** (2016) 012014 doi:10.1088/1757-899X/111/1/012014

Coatings with Reduced Flammability for Linen Fabrics

A Bernava [1], S Reihmane [2]
1-2 Riga Technical University, Faculty of Material Science and Applied Chemistry, Institute of Polymer Materials
P. Valdena Street 3, Riga, LV-1048, Latvia

1 E-mail: aina.bernava@inbox.lv
2 E-mail: reihmane@ktf.rtu.lv

Abstract. The aim of this research is the investigation of burning characteristics and abrasion resistance of the Latvian local raw and bleached linen fabrics and the same fabrics with continuous anti flammable coating, containing commercial printing paste Printperfect EX-TS, pigment Bezaprint GELB (both of the producer CHT BEZEMA) and antimony trioxide (Sb_2O_3) as the flame-retardant additive. For the coating of the linen fabrics the flat screen method was used. After the coating 5 washing cycles were applied to the half of the fabrics. The burning test of the raw and the bleached linen fabrics and those coated with commercial printing paste and antimony trioxide, as well as the washed fabrics was accomplished. The results of the test shove, that the raw and the bleached linen fabrics burn completely. The flame extinction was observed for coated fabrics and coated washed fabrics. The abrasion resistance of coated fabrics depends on thickness of coating.

1. Introduction

Flame protection has become mandatory for personal protective equipment in many branches of industry. Flame protection in combination with the high abrasion–resistance of materials is perspective technology [1]. It is noticeable, that hazards may occur by fires as the combined effect of smoke and toxic gases. Standards and codes of human safety based on recognized general requirements of fire protection are obligatory for lot of textile materials used in building, transport, construction, home and protection garments [2]. All fabrics will burn with some being more combustible than others. The burning rates can be reduced with flame retardants (FR), applied through treatment [3] with chemicals, such as inorganic salts, mix of ammonium polyphosphates, halogen donators, expandable graphite (intumescent) etc. The choice of every individual FR and kind of treatment depends on requirements for produced material. The antimony compounds can be safely used for the textile treatment without causing a risk to human health and the environment [4].

The aim of this research is the investigation of burning characteristics of the Latvian local linen (raw and bleached) fabrics without and with continuous coating, containing commercial printing paste Printperfect EX-TS, pigment Bezaprint GELB and antimony trioxide (Sb_2O_3) as the flame-retardant additive. The manual flat screen method for coating of fabrics was used. The fabrics burning characteristics were tested with the vertical and the horizontal flame applying method, before and after 5 washing cycles. The abrasion resistance was determined on Taber Rotary Platform abrader.

Corresponding author: aina.bernava@inbox.lv

IOP Conf. Series: Materials Science and Engineering **111** (2016) 012014 doi:10.1088/1757-899X/111/1/012014

2. Materials and Methods

2.1. Material.
The Latvian local [Company Larelini Ltd.] plain weave 100% raw and bleached linen fabrics were used for experiment. Before the coating the fabrics were pre-washed in distilled water solution (liquor ratio M 30) of washing agent Felosan NOF (2 g/l) [CHT R Beitlich GMBH] at 100^{0}C for 60 min with following rinse in cold / warm water. The raw linen fabric – sample designation R0 fabric surface density 178 g/m^2, thickness 0.24 mm (Fig. 1A); the bleached linen fabric – sample designation B0 fabric surface density 190 g/m^2, thickness 0.23 mm (Fig.1B).

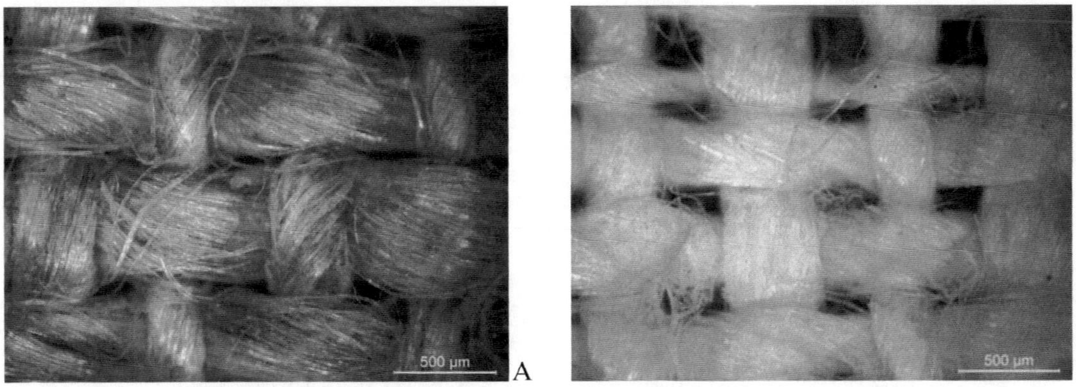

Figure 1. A- raw linen fabric (R0), B- bleached linen fabric (B0)

2.2. Coating.
Coating composition on base of poliacrylate dispersion swelling printing paste Printperfect EX-TS [commercial producer CHT BEZEMA] with flame retardant additive Sb_2O_3 (proportion 9:1) and pigment Bezaprint Gelb 5G (2 mass % of paste) was mixed manually. The flat screen method for coating of the raw linen (sample designation R/C) and the bleached linen (sample designation B/C) fabrics were used. Drying of coated samples at 100^0 C for 7±1 min and the thermal treating at 150^0 C for 5±1 min was applied.

2.3. Microscopy
The linen fabric surface investigation with the image of optical microscopy [Microscope Leica Microsystems, GmbH, magnification 50] was carried out.

2.4. Whiteness level
Whiteness level of the linen fabric surface was determined with Rhopoint Novo-Shade Duo 45°/0 reflectometer [Rhopoint Instrumentation Ltd, UK]. Preparing and testing of samples according to user manual [5] was performed.

2.5. *Washing of coated fabrics*
For the half of coated raw linen (sample designation R/C/W) and coated bleached linen (sample designation B/C/W) fabrics 5 washing cycles in distilled water solution (M 30) of sodium carbonate (3 g/l) and washing agent Felosan NOF (5 g/l) at 40^{0}C for 15 min with following rinse in cold / warm water, according the ISO 105-C10:2006 was used.

2.6. Testing of abrasive resistance
The abrasion resistance of untreated and coated samples (Ø10.2mm) on Taber Rotary Platform abrader, according LVS EN ISO 5470-1:2001 was determined.

2.7. Burning tests

The fabric samples (8 for vertical and 6 for horizontal test) were used according LVS EN ISO 15025:2003. The burning tests were realized in Latvian Centre of Certification (LATSERT).

3. Results and Discussion

The continuous coating on the right side of fabrics was made with manual flat screen method. The Fig.1 and table 1 shove the difference of used base fabrics. The raw fabric low whiteness (32, 23 %) testifies, that fabric refining only with prewashing is not completely. It causes the difference of coating thickness (35%) made with identical technology on raw and bleached textiles.

Table 1 Some characteristics of uncoated and coated linen fabrics.

Designation of sample	R0	B0	R/C	B/C
Thickness of sample, mm	0.24	0.23	0.69	0.45
Whiteness of sample, %	32.23	78.72	-	-
Abrasion resistance of sample, cycles	25	24	1080	660

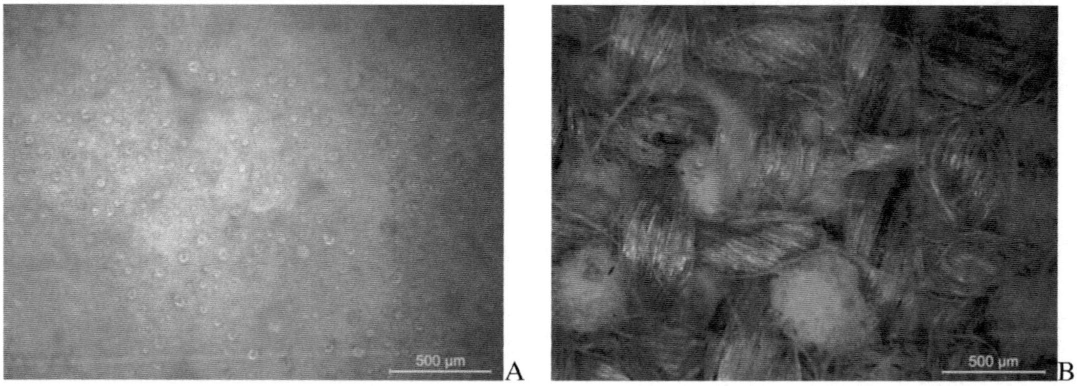

Figure 2. Coated raw linen samples (R/C); right side (A) left side (B).

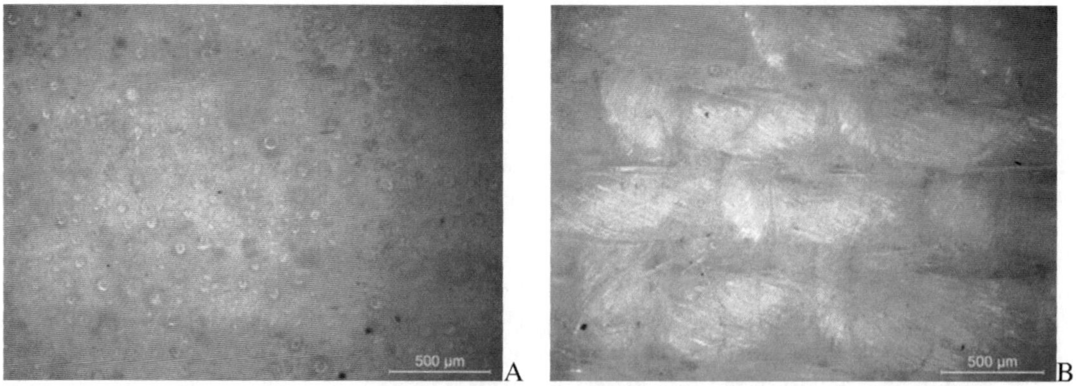

Figure 3. Coated bleached linen samples (R/B); right side (A) left side (B).

The partial coating composition passing through from the right side to the left side of the fabric is observed (Fig. 2B, 3B). There is practically no difference of abrasion resistance for uncoated (25 cycles for R0 sample and 24 cycles for B0 sample) base fabrics. The abrasion resistance of coated materials increases significantly – 43 times for R/C and 28 times for B/C samples. The coating thickness influence on abrasion resistance is observed (Table 1) – the growth of sample thickness for 35 % (R/C in comparison with B/C) causes increase of abrasion resistance for 39 %.

The main task of investigation is decrease of burning characteristics of coated materials on base of linen fabrics. The results of burning tests are presented in table 2.

Table 2 Burning tests results of uncoated and coated linen fabrics.

Designation of sample	R0	B0	R/C	B/C	R/C/W	B/C/W
Flame spreading time till top of the sample, s						
Horizontal test	8-9	10-11	27-30 (right side) 27-29 (left side)	20-21s (right side), 23-24s (left side)	23- 24 (right side), 24-25 (left side)	Don't reaches (right side) 25-26 (left side)
Vertical test	8-9	8-9	10-13 (both sides)	10-11 (both sides)	8-9 (right side), 9-10 (left side)	9-11 (both sides)
After burning time, s						
Horizontal test	Burns complete ly 50-60	Burns complete ly 65-70	Flame extinction 100-115 (right side), 120-130 (left side)	Flame extinction 25-27 (right side), 91-95 (left side)	Flame extinction 60-65 (right side), 70-80 (left side)	Flame extinction 19-22 (right side), 93-95 (left side)
Vertical test	Burns complete ly 40-45	Burns complete ly 40-42	Burns completely 60-80 (right side), 65-67 (left side)	Burns completely 70-73 (right side), 62-68 (left side)	Burns completely 70-71 (right side), 65-68 (left side)	Burns completely 67-69 (right side), 66-70 (left side)
After glowing time, s						
Horizontal test	77-82	75-80	150-177 (right side), 175-190 (left side)	80-85 (right side), 135-142 (left side)	105-133 (right side), 136-140, (left side)	80-83 (right side), 154-160 (left side)
Vertical test	55-60	65-75	182-186 (right side) 170-175 (left side)	160-164 (right side), 150-161 (left side)	175-180 (right side), 177-180 (left side)	157-160 (right side), 154-163 (left side)
Formation of debris						
Horizontal test	Yes	Yes	Yes	No (right side) Yes (left side)	Yes	No (right side); yes (left side)
Vertical test	Yes	Yes	Yes	Yes	Yes	Yes

The characteristic of burning was tested with the horizontal and the vertical method (Table2). The obtained results differ with used method – in all cases the higher results are obtained with horizontal test method.

The raw and the bleached linen fabrics before coating burn completely with the both used methods for 40-70 s. The use of flame resistant coating increases the materials resistance to burning. As expected the right side of samples with uninterrupted coating (Fig. 2, Fig. 3) is more resistant to flame as the left side. In horizontal test the extinction of flame is observed. The lower after burning and after glowing time is obtained for samples with bleached linen fabric base despite the lower coating thickness of samples in comparison with coated samples on base of the raw linen. Formation of thermal degradation product – debris is observed with both testing methods practically in all cases (exception – the horizontal test for the right side of bleached linen coated and coated washed samples). The influence of washing of coated textiles is not significant.

4. Summary

- The raw and the bleached linen fabrics used as the base of coating burn completely,
- The horizontal and the vertical burning test results differ. In the horizontal test the extinction of flame is observed,
- The right side of samples with uninterrupted coating is more resistant to flame as the left side,
- The base fabric selection influences the coating thickness, burning characteristics and abrasion resistance of obtained materials,
- The influence of 5 washing cycles on flame resistance of coated textiles is not significant,
- The investigated coating technology increases the abrasion resistance of examined textile fabrics 28 – 43 times.

Acknowledgments

The authors would like to thank the Docent of Department of Polymer Materials of Riga Technical University Dr. S. Gaidukov for the microscopy of linen fabrics.

References

[1] Schoeller Textile 2015, Permanent flame protection for fabrics, *Technical Textiles* 2, E 116.
[2] A Borisova, S Reihmane, 2012, Flame Retardant Cellulosic Textiles, *Tekstilna Industrija* 3, 5-12.
[3] Flame Retardants Workshop. 2015 February. EFRA March 2015, 3- 4.
[4] Safety of antimony compounds for the textile industry, 2014 May, *International Antimony Association*(VZW), http://www.antimony.com.
[5] Rhopoint Novo-Shade Duo tm 45°/0 Reflectometer. User Manual. 2005. www.rhopointinstruments.com

Microwave Synthesis Of Polyols For Urethane Materials

V Zeltins [1], U Cabulis [1], A Abolins [1], S Gaidukovs [2]

[1]Latvian State Institute of Wood Chemistry, Riga, Latvia
[2]Riga Technical University, Faculty of Material Science and Applied Chemistry, Institute of Polymer Materials, P. Valdena Street 3, Riga, LV-1048, Latvia.

E-mail: sergejs.gaidukovs@rtu.lv

Abstract. The preparation of polyurethane (PU) polyols processed with the use of microwave (MW) irradiation heating polyols is reported. The PU polyols were consisted from bio-based rapeseed oil (RO) and recycled polyethylene terephthalate (PET) components are reported. Obtained MW polyols were compared to the conventional PU polyols synthesized by traditional heating in oil bath. The polyols have the similar properties. The obtained MW assisted polyols were characterized by Fourier transform infrared spectroscopy (FT-IR) measurements. Acid number, hydroxyl value, moisture content and viscosity of the MW synthesized polyols were measured. The obtained polyols consisted from about 34% RO and 23% PET components. The advantages of the very fast reaction rate for preparation of PU polyols, by using MW instead of the conventional heating process, are shown.

1. Introduction

All types of thermally driven chemical reactions can be accelerated by the use of microwaves (MW). From the early 1980-ties, many investigations have been performed in the field of MW synthesis of different organic substances. It is possible to decrease significantly the synthesis reaction time from hours to minutes. Arrhenius equation (1) testifies that very low activation energy should be applied to start the chemical reaction [1].

$$k = Ae^{-E_a/RT} \tag{1}$$

Recently, natural raw components for polyurethane (PU) preparation have attracted considerable interest because of the environmental concerns [2]. Rapeseed oil (RO) can be chosen as bio-based and renewable feedstock to produce polyol components for PU materials manufacturing [3]. RO does not contain necessary hydroxyl groups and cannot react with the isocyanate components, but it can be chemically converted to the PU polyol. For the chemical modification are used reactions of ozonolysis [4], epoxidation [5], hydroformylation [6], esterification [7] and amidization [8].

The utilization of recycled polyethylene terephthalate (PET) for production of PU, has become very popular in the last years [9]. The depolymerization products of PET can be successfully utilized in the production of PU polyester polyols, which are also successfully used for preparation of the PU foams [10]. The glycerol (GL) and adipic acid (ADA) are generally used to improve the branching and functionality of the polyol [11]. Researchers also

report that MW irradiation can be utilized for the PET depolymerization and further preparation of the PU polyols [12, 13].
The main aim of the present paper is connected with synthesis of polyols for urethane materials preparations. The synthesized PU polyols consisted from renewable and recycled components. The clear PET bottle waste was depolymerized by the glycolysis reaction with diethylene glycol (DEG) under MW irradiation conditions. The MW polyols were compared with conventionally prepared polyols. The results of the both synthesis technologies were compared and discussed.

2. Experimental

The MW assisted synthesis was accomplished on the lab. scale 1-liter full glass reactor equipped with MW generator, mechanical stirrer, argon flux and condenser. The photo of the experimental laboratory glass reactor equipped with MW generator is shown in Fig.1. The MW generator frequency was 2450 MHz, power – 800 W. Considering the precise temperature control of the reaction bulk, the applied power for the synthesis was only 60%.
PET flakes of clear grade were granted from PET Baltija, Latvia. RO (I_2=117 I2 mg/100 g) was supplied from Iecavnieks &Co, Latvia. The three different polyols were prepared with MW assisted synthesis process. The first MW polyol was RO/TEA, synthesized at 170 °C; RO and TEA molar ration was 1 M: 2.9 M. The second polyol was RO/PET. After the synthesis of RO/TEA, the 2 M of the recycled PET flakes and DEG were introduced to the reaction mixture. The temperature was increased up to 230 -240 °C; reaction time was about 40 min. The third polyol was similar to the second, but GL and ADA additives were introduced to the reaction mixture to enhance the functionality of the product. GL and ADA content was equal to 1 wt.%.
For the comparison purposes, the same polyols were also prepared at 170– 240 °C in an oil bath, under argon flux and with mechanical stirrer. The content of the raw components for this series was the same as for the MW synthesis.
The Fourier transform infra-red spectra measurements (FTIR) were performed for chemical structure identification. The FTIR measurements were made on a Perkin-Elmer spectrometer Spectrum One FTIR Spectrometer.

3. Results and discussions

At the beginning, the advantages of the MW heating technology application under the conventional heating process can be defined by the investigation of the synthesis processes time kinetics. MW assisted synthesis process of PU polyols was compared with the same PU polyol prepared with conventional heating process. Table 1 has summarized the synthesis time parameters and characteristic properties of the processed polyols.
The three parameters were used to characterize the process kinetics: t_1 – time when the necessary synthesis temperature is achieved (relating to the process, the temperature was 170 and 240 °C); t_2– time when homogeneous mixture is achieved; t_3 –synthesis full time. It should be stressed that the necessary temperature to start PET glycolysis and polyol synthesis reaction can be achieved after 2-4 minutes of MW irradiation heating. For the conventional process t_1 was about 2-3 hours for all polyols. The average MW synthesis full time t_3 for the polyols was about 6 minutes for RO/TEA, and about 40 minutes for RO/PET and

RO/ADA/GL polyols. It is clear that the MW process rate were significantly higher, than, conventional heating synthesis of the same polyols. Depending from the polyol, the MW process time was about 4-40 minutes consuming much lower energy, than conventional electrical heating, which was about 2-3 hours. More rapid MW heating is connected with high dipole moments of the used raw molecules [1]. In the conventional process, the reaction mixture is in contact with the reactor glass walls, which are heated first with the use of conducting medium. Then heat conductance through the oil medium is of very low efficiency [5]. This can be used to explain the difference between the rates of the two heating processes.

Fig.1. Experimental MW assisted laboratory synthesis facility

The obtained polyols' OH values are from 350 till 400 mg KOH/g. All processed polyols have low acid numbers, which are less than 7 mg KOH/g. The received values are suitable for the production of rigid PU foams. Water content in all polyols is less than 0.35% which is also acceptable for obtaining the PU foams.

The PU polyols synthesis has led to the formation of a viscous liquids, which did not solidify on cooling. The color of the polyols was from yellow to brown. The color of the polyols does not depend from the applied heating technology. The MW assisted polyols` viscosity, reported in Table 1, was in the range from 160 to 1250 mPas for both polyols series. The polyols RO/TEA is characterized as a low molecular compound [8]. RO/PET and RO/ADA/GL polyols are more oligomeric [14]. GL/ADA addition in the synthesis modifies the chemical structure of the polyol and decreases the viscosity of the polyol. The obtained viscosity values are also appropriate for manufacture of the PU foams [15].

Table 1. Property comparison of MW assisted synthesis polyols with conventional process polyols

No	Property	MW assisted polyols			Conventionally heated polyols		
		RO/TEA;	RO/PET	RO/GL/ADA	RO/TEA;	RO/PET	RO/GL/ADA
1	t_1, min	2	4	4	120	235	75
2	t_2, min	6	40	40	60	60	75
3	t_3, min	6	50	90	120	120	135
4	Color	yellow	brown	brown	yellow	brown	brown
5	Hydroxyl number, mgKOH/g	360	348	408	363	235	407
6	Acid number, mgKOH/g	n.a.	6.5	3.9	2.5	3.2	4.2
7	Viscosity, mPa s	n.a.	810	630	160	1250	725
8	Water content, wt.%	0.19	0.36	0.12	0.12	0.32	0.14

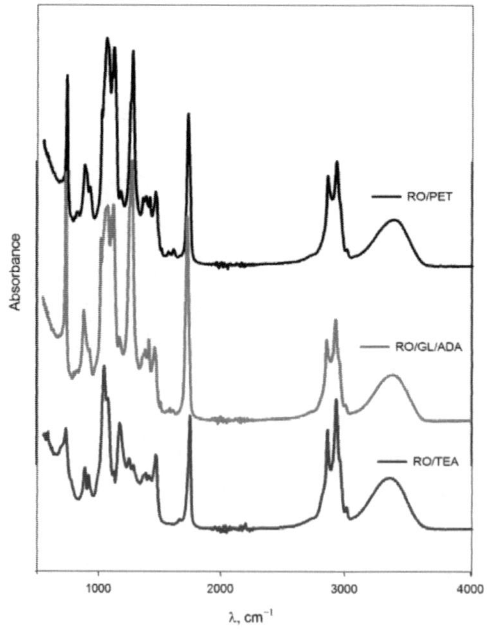

Fig.2. FTIR-ATR spectra for MW assisted polyols RO/PET, RO/GL/ADA, and RO/TEA

The obtained FTIR spectra of the polyols prepared by the both heating technologies are very similar. The FTIR spectra of the MW process are presented in Fig.2. The spectra of RO/TEA is similar to the previously discussed in the literature [16]. The spectra of the MW synthesized RO/PET and RO/GL/ADA polyols confirmed the PET glycolysis with DEG. FTIR spectra testified very close chemical structure of the polyols. The presence of recycled PET component in the polyol is mainly indicated by the absorption band intensities of aromatic C=C vibrations at 1578, 1505 and 1410 cm^{-1}, the absorption band of bending and stretching vibrations of aromatic C–H at 875 and 730 cm^{-1} and the stretching vibration band of aromatic C–O carbonyl group at 1265 and 1120 cm^{-1}. The broad band at 3400 cm^{-1} characterizes the free hydroxyl –OH groups in all polyols. There is also an important absorption band at 1710 cm^{-1}, which corresponds to the carbonyl C=O bonds in ester groups.

It should be also mentioned that the synthesized polyols consisted from 34% renewable (RO) and 23% recycled (PET) components; they combined the advantages of aromatic and aliphatic structure blocks in the molecule backbone.

Additionally, it can be noted that the rigid PU foams preparation was also tested in the lab. The MW obtained polyols were mixed with different additives to obtain general-purpose polyol compositions, which are suitable for preparation of rigid PU foams with thermal insulation properties. The PU foams prepared from MW assisted polyols have very similar structure and properties to the PU foams processed from conventional ones.

4. Conclusions

The utilization of MW for synthesis of PU polyols was testified experimentally in the lab. MW polyols have the similar properties as the traditionally synthesized polyols. Microwave assisted synthesis took place at extremely higher rates as compared to the conventional oil bath heating technology. Acid number, hydroxyl value, moisture content and viscosity of the MW polyols have appropriate values to prepare rigid PU foams. The synthesized PU polyols consisted from about 34% renewable (RO) and 23% recycled (PET) components.

5. References

[1] Hayes B 2002 *Microwave synthesis: Chemistry at the speed of light* / CEM Publishing)

[2] Petrovic Z S 2008 Polyurethanes from vegetable oils *Polym. Rev. (Philadelphia, PA, U. S.)* **48** 109-55

[3] Meier M A R, Metzger J O and Schubert U S 2007 Plant oil renewable resources as green alternatives in polymer science *Chemical Society Reviews* **36** 1788-802

[4] De Souza V H R, Silva S A, Ramos L P and Zawadzki S F 2012 Synthesis and characterization of polyols derived from corn oil by epoxidation and ozonolysis *JAOCS, Journal of the American Oil Chemists' Society* **89** 1723-31

[5] Zhang C, Ding R and Kessler M R 2014 Reduction of epoxidized vegetable oils: A novel method to prepare bio-based polyols for polyurethanes *Macromol. Rapid Commun.* **35** 1068-74

[6] Petrović Z S, Cvetković I, Milić J, Hong D and Javni I 2012 Hyperbranched polyols from hydroformylated methyl soyate *J. Appl. Polym. Sci.* **125** 2920-8

[7] Valero M F and Gonzalez A 2012 Polyurethane adhesive system from castor oil modified by a transesterification reaction *Journal of Elastomers and Plastics* **44** 433-42

IOP Conf. Series: Materials Science and Engineering **111** (2016) 012015 doi:10.1088/1757-899X/111/1/012015

[8] Kirpluks M, Cabulis U, Kurańska M and Prociak A 2013 Three different approaches for polyol synthesis from rapeseed oil. *Key Eng. Mater.,* **559** 69-74
[9] Dutt K and Soni R K 2013 A review on synthesis of value added products from polyethylene terephthalate (PET) waste *Polymer Science - Series B* **55** 430-52
[10] Vaidya U R and Nadkarni V M 1989 Polyester polyols from glycolyzed PET waste: effect of glycol type on kinetics of polyesterification *J. Appl. Polym. Sci.* **38** 1179-90
[11] Kathalewar M, Dhopatkar N, Pacharane B, Sabnis A, Raut P and Bhave V 2013 Chemical recycling of PET using neopentyl glycol: Reaction kinetics and preparation of polyurethane coatings *Prog. Org. Coat.* **76** 147-56
[12] Chaudhary S, Surekha P, Kumar D, Rajagopal C and Roy P K 2013 - Microwave assisted glycolysis of poly(ethylene terepthalate) for preparation of polyester polyols - **129** - 2788
[13] George N and Kurian T 2014 Recent Developments in the Chemical Recycling of Postconsumer Poly(ethylene terephthalate) Waste *Ind. Eng. Chem. Res.* - **53** 14198
[14] Patel M R, Patel J V and Sinha V K 2005 Polymeric precursors from PET waste and their application in polyurethane coatings *Polym. Degrad. Stab.* **90** 111-5
[15] Ashida K 2006 *Polyurethane and Related Foams*: CRC Press)
[16] Fridrihsone A, Stirna U, Lazdiņa B, Misane M and Vilsone D 2013 - Characterization of polyurethane networks structure and properties based on rapeseed oil derived polyol - **49** 1214

IOP Conf. Series: Materials Science and Engineering **111** (2016) 012016 doi:10.1088/1757-899X/111/1/012016

Preparation and characterization of hot-pressed Li^+ ion conducting PEO composite electrolytes

A Jurkane and S Gaidukov
Institute of Polymer Materials, Faculty of Material Science and Applied Chemistry, Riga Technical University, P.Valdena Street 3, Riga, LV-1048, Latvia.

E-mail: alex.kijanica@gmail.com, sergej.gaidukov@gmail.com

Preparation and characterization of hot-pressed Li^+ ion conducting Polyethylene oxide (PEO) composite electrolytes are discussed in this paper. Lithium trifluoromethanesulfonate with linear formula CF_3SO_3Li is selected as Li^+ ion source for polymer composite electrolyte processing. Polyethylene oxide as matrix for polymer electrolyte is selected, because of its cation solvatation and salt complexes formation partibility. Nano-sized fillers such as carbon nanotube (CNT) and graphene (GR) are introduced to polymeric composite to improve PEO electrolyte mechanical properties. The composites were prepared from PEO, CF_3SO_3Li, CNT and GR by solution casting and by hot-pressing method. Influence of CF_3SO_3Li, CNT and GR concentrations on the electrical, thermal and mechanical properties of PEO composite is investigated and discussed. The additives addition to the PEO matrix improves the electrical, thermal and mechanical properties of the composite. The crystallinity of the PEO decreases.

1. Introduction

The term "battery" describes a device that converts the chemical energy contained in its active materials directly into electric energy through the electrolyte, which provides the essential ionic transfer [1-3]. Wide range of film-like solid polymer electrolyte designs are already utilized in batteries, mostly in portative electronics because of light-weight, non-toxicity, non-combustibility, easy forming, mechanical stability [1,3-9]. For the solid polymer electrolyte system it is important that polymer has a strong coordination to the cations. Polyethylene oxide (PEO) have no double bonds and have a "plenty" of space between its oxygen groups, which is preferable for many different types of cation solvatation and salt complexes formation [3,8-10]. Nuclear magnetic resonance spectroscopy results of PEO-Lithium salt systems show that conductivity is dominated by the movement of lithium ion, i.e., the lithium transport number is 1.0 [11]. Melting temperature "T_m" of the polymers is one of the most important technological parameters, because all processes and practical application of ready material are highly affected by it [12]. Differential scanning calorimetry (DSC) results indicate that addition of Lithium salt decrease important melting values, such as degree of crystallinity and melting temperatures [2,13]. It can be concluded, that increasing the amorphous phase amount and mobility of Lithium cations could increase the ionic conductivity of polymer electrolyte [1,10]. At the same time, the limit of conductivity for applications is 10^{-5} S/cm (in the range between 20°C and 100°C). The limit for conductivity to amorphous phase is its glass transition temperature, but no higher than it. Some authors claim, that ionic conductivity in solid PEO-Lithium salt system reached averaged maximum 10^{-6} S/cm [2,14,15]. However, ionic conductivity of PEO-alkali metal salt system was reached 10^{-7} S/cm at room temperature. PEO-Lithium salt system were revealed to be amorphous when the molecular weight was in range 300-750 g/mol, but system that was obtained with PEO higher than 1000 g/mol - partly crystallized [16]. This phenomenon is based in decrease of crystallite size, which occurs at higher PEO molecular weight, although it cannot explain the different activation energies in the systems [17]. PEO-Lithium salt systems show highest conductivity with PEO molecular weight value lower than 300 000 g/mol.

Lithium trifluoromethanesulfonate (LiTrifl) with linear formula CF_3SO_3Li is widely used as ion source and plasticizer [2,3,8-11,13-19]. The Lithium ion is the stronger Lewis base of the group. It accepts the lone pairs on oxygen the quickest and easiest and has the highest relative exposed positive charge. Hence it will attract water, making the salts hygroscopic and therefore more flexible [20]. Hydrates still exist in Lithium salts after 12 h of drying at 300°C [21,22]. LiTrifl start its decomposition at 290°-300°C [15,23].

Aim of nano-filler addition is the reducing the crystallizing ability of the polymer without reducing the mechanical properties of the PEO composite. The carbon nanoparticles network-polymer composite exhibits electron tunneling conductivity, which show that the CNTs and GRs network offers a promising application as filler material for creating super conductive composite [15,24-26].

The aim of this paper is to explore LiTrifl salt concentration effect on electrical conductivity and thermal properties of hot-pressed Li^+ ion conducting PEO composite electrolytes.

2. Methodology

2.1. Sample preparations

PEO was chosen as matrix of composite polymer electrolyte. LiTrifl as ion source. As reinforcing additives were chosen: CNT and GR. Properties of substance given above were supplied by manufacturer and are listed in table 1.

Table 1. Material used in sample preparation

Substance	Manufacturer	Substance properties
PEO	*Scientific Polymer Products*	MW, 100 000 g/mol; relative d = 1.21 g/cm^3; Tm = 65°C
LiTrifl	*Fluorochem*	MW 156.01 g/mol
CNT (multiwall)	*Nanocyl™ NC 7000*	relative d = 1.3-2 g/cm^3; E = 1TPA, σ_B =10-60 GPa, ε_B = 10% thermal conductivity, >3000 W/m·K σ =10^6-10^7 Sm/m; average D = 9.5 nm, average L = 1.5 μm
GR	XGnP Sciences	relative d = 2.2 g/cm^3; E(‖)=1GPa; σ_B(‖)=5 GPa thermal conductivity (‖), >3000 W/m·K σ =10^2-10^7 Sm/m; average thickness = 7.5 nm average surface area = 120...150 m^2/g

6% PEO solution was prepared by dissolving PEO in dimethylformamide (DMF) and mixed with magnetic stirrer at 70°C (20 min). LiTrifl, CNT and GR were dispersed in DMF by ultrasound for 5 min. It was added to 6% PEO solution and dispersed with additional ultrasound for 10 minutes. The polymer solution was casted as film and solvent was allowed to evaporate for 24 h at 45°C with subsequent annealing in vacuum at 45°C for 24 h. Films with average 500 μm thickness were prepared in mold at 90°C (6 min) by heat-pressing equipment. Concentrations of CNT and GR in composite were 0.05, 0.01, 0.25, 0.50 % of dry PEO mass. Two different system of LiTrifl concentration were separately studied, for CNT: 5, 15, 25% and for GR: 0.5, 1.5, 2.5% of dry PEO mass.

2.2. Volume resistivity measurements

The electrical conductivity of the sample was measured by sandwiching the samples between 2 electrodes using "Keithley 6487" picoampermeter. Constant 500V voltage for 60 s at 23°C was applied. Direct current conductivity measurements were made. Root mean square errors for obtained data were calculated, which are less than 5% for all samples with two parallel measurement. The resistivity to the flow of current through a material is known as volume resistivity (R_V). Specific volume resistance (ρ_V) for the thin sample should be calculated using the following equation (1):

$$\rho_V = R_V \cdot 22.9cm^2/h, [\Omega \cdot cm] \quad (1)$$

R_V – measured volume resistance of the sample (V/I), [Ω]; h – thickness of the sample, [cm]; *22.9*– surface area of the electrode, [cm^2].

2.3. Differential Scanning Calorimetry

Thermal analysis was performed on "Mettler Toledo DSC30" equipment at heating and cooling rates of 10°C/min from 20°C to 100°C under a N_2 atmosphere. It is known, that completely (100%) crystalline PEO polymer melting enthalpy ($\Delta H°$) is equal to 213.7 J/g [6,7]. Using the equation (2) from below, we can calculate degree of crystallinity of PEO composite samples:

$$\chi = \Delta H_m / \Delta H_m^o \cdot 100\% \quad (2)$$

χ- degree of crystallinity of sample, [%]; ΔH_m - melting enthalpy of sample, [J/g]; $\Delta H_m°$ - completely (100%) crystalline PEO polymer melting enthalpy, [J/g]

2.4. Mechanical properties

Mechanical testing was performed using a computer-controlled micro-force tester "ZWICK BDO-FB 020TN". The composite film sample in the form of film "dogbone" strip (20mm x 5 mm). The speed for the elongation test was maintained at 5 mm/min at 23°C.

3. Results and discussions

3.1. Conductivity measurements

Calculated electrical conductivity from specific volume resistance is summarized in figure 1, 2 and 3.

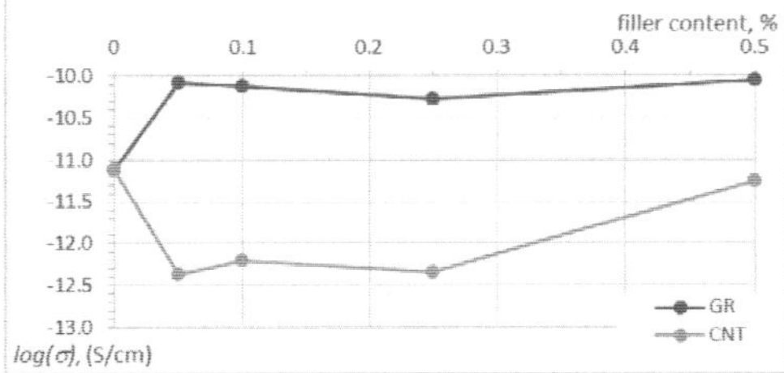

Figure 1. Effect of CNT and GR content on the conductivity of the PEO/x%CNT/0%LiTrifl and PEO/x%GR/0%LiTrifl composite electrolyte (x=0, 0.05, 0.1, 0.25, 0.5) at 23 °C

Conductivity results (figure 1) of the PEO/CNT and PEO/GR composite show, that addition of 0.05% GR increase conductivity of pure PEO from $7.8 \cdot 10^{-12}$ S/cm to $82.4 \cdot 10^{-12}$ S/cm. It can be concluded, that addition of only 0.05% GR increase conductivity of pure PEO by 10 times. At the same time, the opposite situation is with addition of 0.05% CNT, because conductivity decreased by 18 times from $7.8 \cdot 10^{-12}$ S/cm to $0.42 \cdot 10^{-12}$ S/cm. This behaviour can be explain by crystallinity degree increase and decrease of free Li^+ ion mobility. It has been noticed in figure 1 that the conductivity decreases with the increasing content of filler in the range of 0.1% to 0.25%. However, the conductivity start increase when the percentage of CNT and GR keeps on increasing.

Conductivity results (figure 2) of the PEO/CNT/LiTrifl composites show, that addition of LiTrifl increase electrical conductivity in all PEO/CNT/LiTrifl composites cases. Addition of 5% LiTrifl increase electrical conductivity of pure PEO from $7.8 \cdot 10^{-12}$ S/cm to $54.3 \cdot 10^{-12}$ S/cm (by 7 times), addition of 15% LiTrifl to $40.9 \cdot 10^{-12}$ S/cm (by 5 times) and addition of 25% LiTrifl to $32.8 \cdot 10^{-12}$ S/cm (by 4 times).

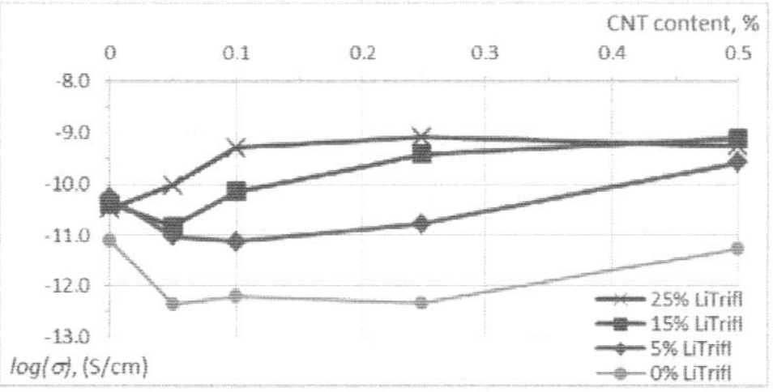

Figure 2. Effect of LiTrifl content on the conductivity of the PEO/x%CNT/n%LiTrifl composite electrolyte (x=0, 0.05, 0.1, 0.25, 0.5; n=0, 5, 15, 25) at 23 °C

It has been noticed in Fig.2 that the conductivity of PEO/CNT/LiTrifl composites raises with the increasing content of CNT in the range of 0.1% to 0.25%. As it was mentioned in literature [27,28], the presence of CNT and GR nanoparticles enhances the electric conductivity substantially. It can be assumed that increase in the conductivity occurs as a result of filler addition, possible amorphous phase increase free Li^+ ion higher mobility.

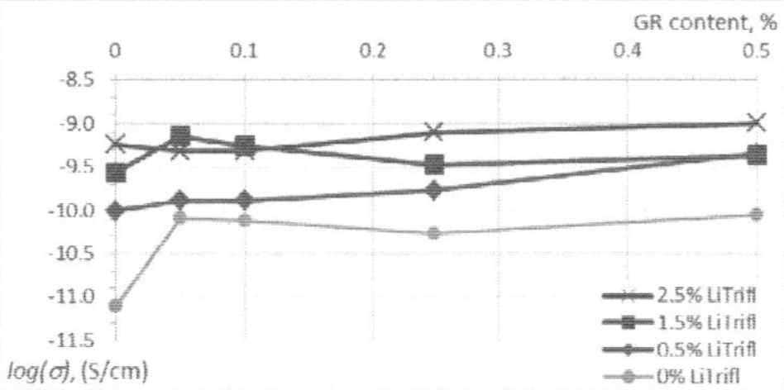

Figure 3. Effect of LiTrifl content on the conductivity of the PEO/x%GR/n%LiTrifl composite electrolyte (x=0, 0.05, 0.1, 0.25, 0.5; n=0, 0.5, 1.5, 2.5) at 23 °C

Similar behavior of electrical conductivity properties of PEO/GR/LiTrifl (figure 3) was observed before in PEO/CNT/LiTrifl composites. Addition of LiTrifl increase electrical conductivity in all PEO/GR/LiTrifl composites cases. Addition of 0.5% LiTrifl increase electrical conductivity of pure PEO from $7.8\cdot10^{-12}$ S/cm to $99.3\cdot10^{-12}$ S/cm (by 13 times), addition of 1.5% LiTrifl to $273\cdot10^{-12}$ S/cm (by 35 times) and addition of 2.5% LiTrifl to $571\cdot10^{-12}$ S/cm (by 73 times). Observed conduction behaviour could be attributed to higher contact surface of GR. It has been noticed in figure 3 that conductivity of PEO/CNT/LiTrifl with 0.5% and 2.5% LiTrifl content raises with the increasing content of CNT in the range of 0.05% to 0.5%, but decreses with 1.5% LiTrifl content in the range of 0.05% to 0.25%, however further addition of salt raise electrical conductivity.

3.2. DSC results

Results obtained in DSC are summarized in figure 4, 5 and 6. It consists of melting temperatures and degree of crystallinities values. We have tested the lowest and the highest content of the CNT and GR filler to see the main characteristic changes in polymer composite electrolite thermal properties. Conductivity measurements also concluded the provisional relative changes.

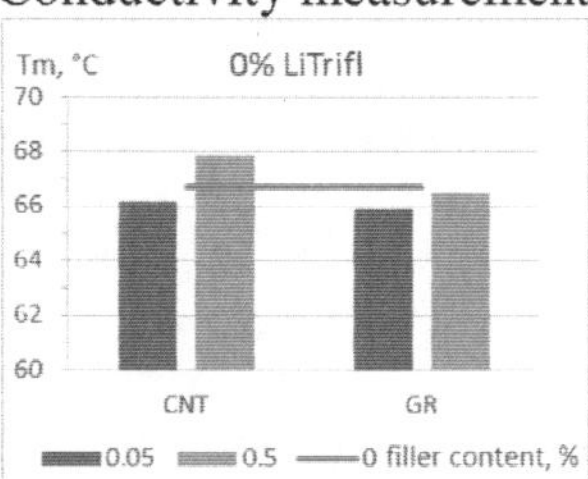

Figure 4. Effect of CNT and GR content on the melting temperature of the PEO/x%CNT/0%LiTrifl and PEO/x%GR/0%LiTrifl composite electrolyte (x=0, 0.5, 0.05)

Initially, it was found that our PEO reached 80% of the theoretically possible crystallinity (in figure 4, 5 and 6 illustrated as straight line). Figure 4 results show, that addition of only 0.05% CNT or GR to pure

PEO, decrease melting temperature from 66.7°C by 0.5°C and 0.8°C respectively. But addition of 0.5% CNT to PEO increase its melting temperature by 1.2°C, therefore electrical conductivity in this case decreased.

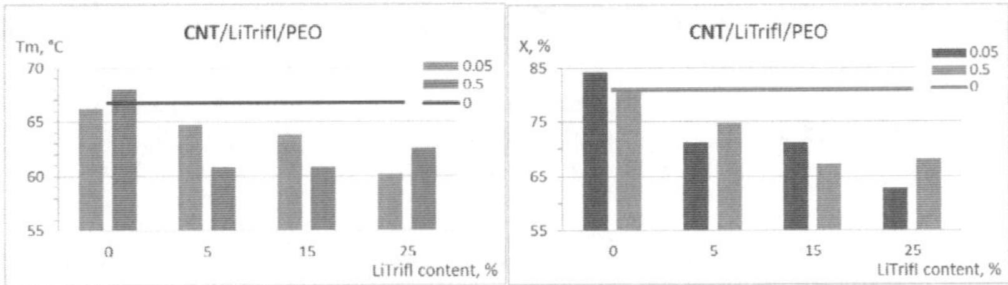

Figure 5. Effect of CNT content on the melting temperature and crystallinity of the PEO/x%CNT/n%LiTrifl composite electrolyte (x=0, 0.5, 0.05; n=0, 5, 15, 25)

Figure 5 and figure 6 results show, that addition of LiTrifl decrease melting temperature and crystallinity in both of PEO/CNT/LiTrifl and PEO/GR/LiTrifl composite electrolyte systems. It should be noted, that addition of LiTrifl decrease melting temperature of PEO/0.05%CNT/LiTrifl composite from 66.2°C to 60.2°C, but addition of LiTrifl to PEO/0. 5%CNT/LiTrifl decreased Tm only to 60.8°C (figure 5), further addition of LiTrifl increase melting temperature. From figure 5 the minimal melting temperature corresponds to PEO/0.05%CNT/25%LiTrifl.

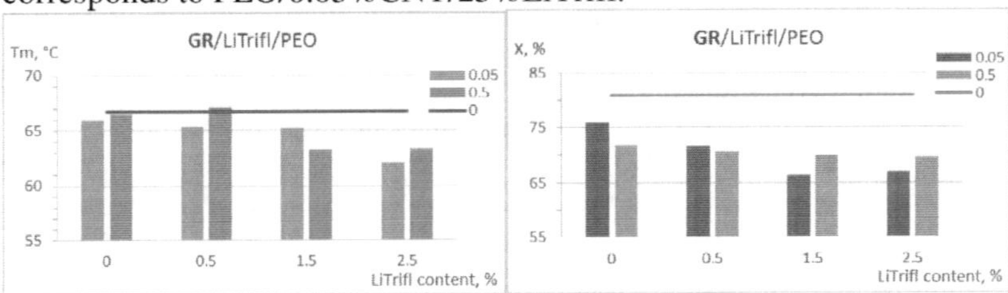

Figure 6. Effect of GR content on the melting temperature and crystallinity of the PEO/x%GR/n%LiTrifl composite electrolyte (x=0, 0.5, 0.05; n=0, 0.5, 1.5, 2.5)

Meanwhile, it also can be observed in the figure 6, that addition of LiTrifl decrease melting temperature of PEO/0.05%GR/LiTrifl composite from 65.9°C to 62.1°C, but addition of salt to PEO/0.5%GR/LiTrifl decrease it only to 63.4°C. Bulk PEO melting temperature is lower than PEO film melting temperature. The difference between bulk (T_m=65°C) and film (T_m=66.7°C) PEO melting temperatures is equal to 1.7°C. We can suggest that this difference exist because of free volume theory and kinetic theory for structural relaxation, where dynamic displacements in film are higher than in bulk polymer [29,30].

3.3. Tensile strength properties

Concluding, we have tested tensile strength properties of PEO composite solid electrolyte samples.

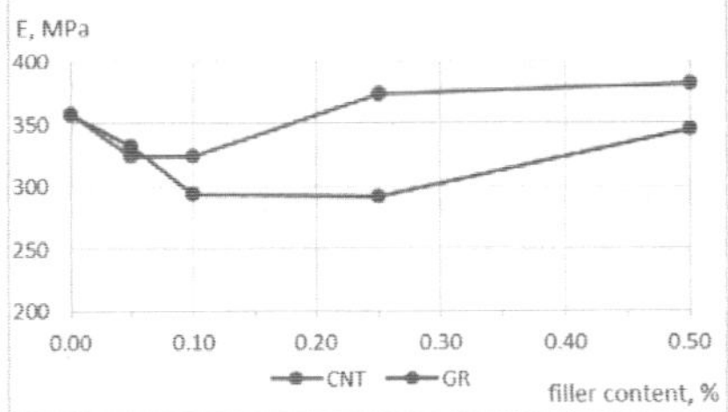

Figure 7. Effect of CNT and GR content on the Young's module of the PEO/x%CNT/0%LiTrifl and PEO/x%GR/0%LiTrifl composite electrolyte (x=0, 0.05, 0.1, 0.25, 0.5)

Figure 7 results show, that pure PEO Young's module is 357 MPa, but addition of CNT or GR reduce it up to 0.1% nano-filler content, further addition of CNT or GR increase module values. It is observed, that composite Young's module values is higher in PEO/CNT case than in PEO/GR. In example, in PEO/0.1%GR composite Young's module is equal to 293 MPa, but in PEO/0.1%CNT it is equal to 323

MPa, therefore it increased by 9%. Addition of 0.25% CNT to PEO/CNT samples increase Young's module by 21%.
Figure 8 and figure 9 results show, that the higher amount of LiTrifl in both PEO/CNT/LiTrifl and PEO/GR/LiTrifl composite, the lowest Young's module values obtained. Such a change in module values can be explained by the decreasing in crystallinity and increasing the amorphous phase by Li^+ ion intercalation between polymer chains, therefore samples become more elastic. Figure 8 results show, that addition of LiTrifl to PEO/CNT/LiTrifl decrease Young's module compare to PEO/CNT composite.

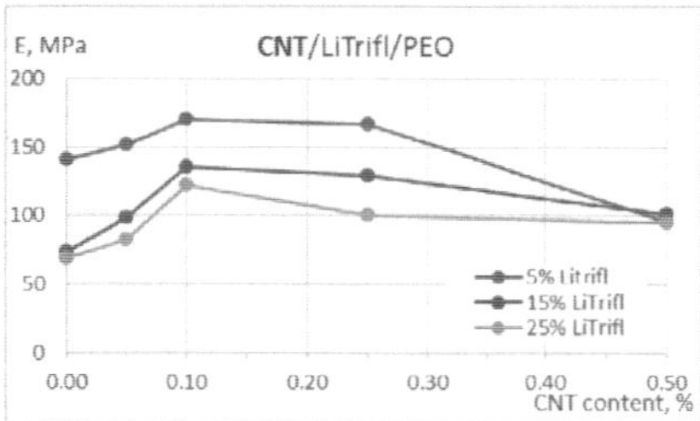

Figure 8. Effect of CNT content on the Young's module of the PEO/x%CNT/n%LiTrifl composite electrolyte
(x=0, 0.05, 0.1, 0.25, 0.5; n=5, 15, 25)

At the same time, figure 9 results show that addition of LiTrifl to PEO/GR/LiTrifl increase Young's module compare to PEO/GR composite electrolyte. As the exception in PEO/GR/LiTrifl system, it was observed, that addition of 0.5% GR to PEO/2.5%LiTrifl composite decrease its Young's module by 68% (figure 9). It was concluded (from figure 9), that the highest Young's module of PEO/GR/LiTrifl system is for PEO/0.5%GR/0.5%LiTrifl and it is equal to 421 MPa; in case of PEO/CNT/LiTrifl system it is for PEO/0.1%CNT/5%LiTrifl and it is equal to 170 MPa.

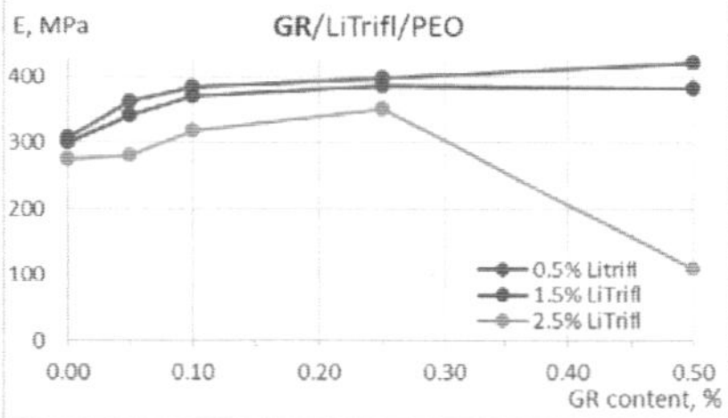

Figure 9. Effect of GR content on the Young's module of the PEO/x%GR/n%LiTrifl composite electrolyte
(x=0, 0.05, 0.1, 0.25, 0.5; n=0.5, 1.5, 2.5)

4. Conclusions
LiTrifl salt is effective as electrical resistance reduction at room temperature. Increasing of the LiTrifl content increase electrical conductivity in PEO/CNT/LiTrifl and PEO/GR/LiTrifl system. The highest electrical conductivity in PEO/CNT/LiTrifl system is for PEO/0.25%CNT/25%LiTrifl and it is equal to $81.8 \cdot 10^{-9}$ S/cm. The highest electrical conductivity in PEO/GR/LiTrifl system is for PEO/0.5%GR/2.5%LiTrifl and it is equal to $1.01 \cdot 10^{-9}$ S/cm. It is evident that the mechanical properties such as Young's module are higher for PEO/GR/LiTrifl system. Addition of LiTrifl as plasticizer to PEO/CNT and PEO/GR system decrease crystallinity and thus decrease melting temperature of the samples.

5. References
[1] Yin Y 2010 An Experimental study on PEO polymer electrolyte based all-solid-state supercapacitor. *Open Access Dissertations* p 440
[2] Stephan M A and Nahm K S 2006 *Polymer (Guildf)* **47** 5952-64
[3] Chew K W, Ng T C and How Z H 2013 *Int. J. Electrochem. Sci.* **8** 6354-64
[4] Di W 2012 *Nanomaterials* **2** 268-274

IOP Conf. Series: Materials Science and Engineering **111** (2016) 012016 doi:10.1088/1757-899X/111/1/012016

[5] Daud F N, Ahmad A and Badri K H 2014 *Int. J. of Poly. Sci.* **2014** 5

[6] Ibrahim S and Johan M R 2012 *Int. J. Electrochem. Sci.* **7** 2596-2615

[7] Ali A M M, Subban R H Y, Bahron H, Winnie T, Latif F and Yahya M Z 2008 *Ionics* **14** 491-500

[8] Chandra An, Chandra Ar and Thakur K 2013 *Indian J. of Pure & Applied Physics*., **51** 44-8

[9] Dey A, Karan S and De K S 2013 *Indian J. of Pure & Applied Physics*. **51** 281-8

[10] Brandell D. 2005 Understanding ionic conductivity in crystalline polymer electrolytes. Acta Universitatis Upsaliensis. *Digital Comprehensive summaries of Uppsala dissertations from the faculty of science and technology* 34. p 63

[11] Gadjourova Z, Andreev Y G, Tunstall D P and Bruce PG 2001 *Nature* **412** 520

[12] Noor S A M et al. 2010 *Natural Sci.* **2** 190-6

[13] Kim H, Choi G, Moon S and Kim S 2003 *J Appl. Electrochem.* **33** 491-6

[14] Gray F M 1997 Fast ion transport in solids. Polymer electrolytes, RSC materials monographs. (Cambridge: The Royal Society of Chemistry) ed: Vashista P, Mundy J N, Shenoy G K (Amsterdam: Elsevier) p 131

[15] Kiyanitsa A, Gaidukov S, Zukulis E and Juhnevica I *2014 CYSENI* (Kaunas, Lithuania) ISSN 1822-7554

[16] Yoichi T, Naomi T and Hiroyuki O 2000 *Electrochimica Acta*. **45** 1285-9

[17] Stoeva Z, Martin-Litas I, Staunton E, Andreev Y G and Bruce PG 2003 *J. Am. Chem. Soc.* **125** 4619

[18] Willgert M 2011 *Eur Polym J* **47** 2732-38

[19] Armand M, Gorecki W and Andreani R 1990 *Proc. 2nd Int. Symp. on Polymer Electrolytes* (London: Elsevier) p 31

[20] Munshi M Z A 1995 *Handbook of Solid State Batteries & Capacitors* (Singapore: World Scientific publishing Co. Pte. Ltd.) p 716

[21] Shumaher I. 1963 Perhlorates. Properties. Moskva: SChI. ed: Genin L.S. p236 (in Russian: Шумахер И.. 1963 *Перхлораты. Свойства*. Москва: ГХИ, под редакцией Генина Л.С., с. 236)

[22] Kurt H S 1974 *J. Phys. Chem. Ref. Data* **3** 481-526

[23] Leijonmarcka S, Carlsonb T and Lindbergha G 2013 *Comp. Sci. and Tech.* **89** 149-157

[24] Ferrari A C 2007 *Solid state comm.* **143** 47-57

[25] Grayfer E D et al. 2011 *J. Mater. Chem.* **21** 3410

[26] Mahmoud W E *Eur.* 2011 *Polym. J.* **47** 1534-40

[27] Chew K W and Tan K W 2011 *Int. J. Electrochem. Sci.* **6** 5792-5801

[28] Yap Y L, You A H, Teo L L and Hapei H 2013 *Int. J. Electrochem. Sci.* **8** 2154-63

[29] Becker J S et al. 2011 *Proc. Natl. Acad. Sci. USA* **108** 977–982

[30] Kawashima K, Inoue R, Kanaya T G et al. 2009 *J of Phys.: Conf. Ser.* **184** 012004

IOP Conf. Series: Materials Science and Engineering **111** (2016) 012017 doi:10.1088/1757-899X/111/1/012017

Modified silyl-terminated polyether polymer blends with bisphenol A diglycidyl ether epoxy for adhesive applications

J Bitenieks[1], R Merijs Meri[1], J Zicans[1], R Berzins[2], J Umbraško[2] and U Rekners[2]
[1]Institute of Polymer Materials, Faculty of Material Science and Applied Chemistry, Riga Technical University, Riga, Latvia
[2]TENACHEM Ltd., TENAX group, Dobele, Latvia

Abstract. Modified silyl-terminated polyether polymer (MS Polymer) was blended with bisphenol A diglycidyl ether (DGEBPA) epoxy at MS Polymer/epoxy ratio from 30/70 to 70/30. MS Polymer/epoxy systems were examined for two-component adhesive formulation with additional fillers. Applicability of the MS Polymer/epoxy system at the ratio of the components 60/40 is demonstrated for the development of adhesive formulation. Rheological analysis of the components A and B shows suitable viscosity values for development of two-component adhesives formulation. Curing dynamics as well as tensile stress-strain properties and Shore A hardness of the chosen adhesive formulation are reasonable for the development of MS Polymer/epoxy type adhesive.

1. Introduction

Despite many advantages of epoxy resins like high strength, chemical and heat resistance, in many applications they lack flexibility. During service under dynamic stress and variable environment this may lead to cracking and even failure of the adhesive joint. To reduce epoxy resin brittleness they can be belended with elastomer toughening agents. At certain amount these elastomer tougheners can improve fracture, elastic and other properties [1 - 3].

In this work modified silyl-terminated polyether polymer (MS Polymer) is used as toughener for epoxy resin. MS Polymer consists of polyether backbone and methoxysilyl chain ends. Advantages of MS Polymers are excellent adhesion to various substrates and low-temperature elasticity (T_g = -60°C) [4]. MS Polymer can be blended with epoxy resins to utilize the advantages of both epoxy resin and MS Polymer leading to flexible products. When epoxy resin is mixed together with MS Polymer they form heterogenic matrix/domain structure which consists of the epoxy domains functioning as hard segments in elastic MS Polymer matrix. As both polymers are not miscible, in order to obtain optimum strength and flexibility, the epoxy resin microphase must be separated into discrete domains during curing by controlling the epoxy resin/MS Polymer ratio. Information from previous studies shows that in the case of elastic adhesives MS Polymer forms the matrix in its blends with epoxy. Maximum strength level is achieved at specific MS Polymer/epoxy ratio [5].

In this research MS Polymer blends with epoxy resin are studied by varying MS Polymer and epoxy ratio from 30 to 70% to find the best MS Polymer/epoxy ratio for development of adhesive formulation.

2. Materials

Materials used for this research are listed in Table 1.

Table 1. Materials for two-component adhesive investigation

Component	Type	Supplier
SAX 350	MS Polymer	Kaneka Belgium
DER 331	Epoxy resin	DOW Chemical
Tib Kat 318	MS Polymer catalyst	TIB Chemicals
Baxxodur EC 130	Epoxy hardener	BASF
AMMO	Aminosilane	Evonik Industries
Omyacarb 1T VA	Coarse filler	Omya AG
Hakuenka CCR-S10	Fine filler	Shiraishi-Omya GmbH
Cabosil TS-720	Thixotropic additive	Cabot Corporation
Irganox 245	Antioxidant	BASF
Tinuvin 328	UV stabiliser	BASF

3. Experimental

MS Polymer was mixed together with bisphenol A diglycidyl ether epoxy resin (DGEBPA) at various MS Polymer/Epoxy weight proportions starting from 30/70 to 70/30 (epoxy hardener was taken into account in epoxy part). Hardener was added in stoichiometric amount to epoxy considering epoxide equivalent weight 190 g/eq and amine hydrogen equivalent weight 56 g/eq. MS Polymer catalyst, aminosilane compatibilizer and water were adjusted to MS Polymer by weight. The formulations for MS Polymer/Epoxy blends are shown in Table 2.

Table 2. Compositions of the investigated MS Polymer/Epoxy blends

Component	wt, %								
SAX 350	30.00	35.00	40.00	45.00	50.00	55.00	60.00	65.00	70.00
Tib Kat 318	1.00	1.17	1.33	1.50	1.67	1.83	2.00	2.17	2.33
DER 331	54.07	50.20	46.34	42.48	38.62	34.76	30.89	27.03	23.17
Baxxodur EC 130	15.93	14.80	13.66	12.52	11.38	10.24	9.11	7.97	6.83
AMMO	1.30	1.52	1.73	1.95	2.17	2.38	2.60	2.82	3.03
H_2O	0.30	0.35	0.40	0.45	0.50	0.55	0.60	0.65	0.70

Next step was to formulate adhesive with additional fillers. Component A and B ratio was decided to be 2:1 to follow typical commercial two-component product ratio guidelines.The formulations for A and B components are presented in Table 3.

Table 3. Formulation of two-component MS Polymer/Epoxy adhesive with ratio 60/40

Formulation part A			Formulation part B		
Component	Type	wt, %	Component	Type	wt, %
SAX 350	MS Polymer	30.00	DER 331	Epoxy resin	15.00
Baxxodur EC 130	Epoxy hardener	4.50	Tib Kat 318	Catalyst	1.00
Irganox 245	Antioxidant	0.30	Irganox 245	Antioxidant	0.15
Tinuvin 328	Stabiliser	0.30	Tinuvin 328	UV stabiliser	0.15
Omyacarb 1T VA	Coarse filler	19.30	Omyacarb 1T VA	Coarse filler	14.70
Hakuenka CCR-S10	Fine filler	9.50	Hakuenka CCR-S10	Fine filler	2.00
Cabosil TS-720	Thixotropic agent	1.50	H_2O		0.30
AMMO	Compatibilizer	1.30			

MS Polymer/Epoxy blends were mixed by using SpeedMixer DAC 150 centrifugal laboratory mixer to obtain homogeneous mixtures of the components. Samples for mechanical characterisation were casted in Teflon moulds and left for curing for determined duration at 23±2°C and ~ 50±5% RH.
Rheological characteristics were measured by using Bohlin CVO 100 rheometer. Instrument was equipped with 20 mm diameter spindle in plate-plate geometry. The gap size between plates was set to 200 μm. Tests at 23±2°C were performed in constant shear rate mode at 0.2, 0.5, 1 and 10 s^{-1}. During the experiment changes in viscosity η were recorded.
Tensile stress-strain characteristics were determined by using Zwick/Roell Z010 universal testing machine. Tests were made according to DIN 53504 at test speed 200 mm/min. Before determination of tensile properties MS Polymer/Epoxy blends with increasing MS Polymer ratio from 30 – 70% were cured for 14 days at 23±2°C and ~ 50±5% RH. MS Polymer/Epoxy adhesive formulation was cured for 1, 7, 14, 21 and 21 days at 23±2°C and ~ 50±5% RH.
Shore A hardness for MS polymer/epoxy blends after 14 days and MS polymer/epoxy adhesive formulation after 1, 7, 14, 21 and 28 days of curing at 23±2°C and ~ 50% RH was measured according to ISO 868 by using Zwick Roell 3114/5 durometer.
Dynamic mechanical thermal analysis (DMTA) of MS Polymer/epoxy adhesive formulation was performed by using Mettler Toledo DMA/SDTA861device. To ensure the sample was sufficiently cured prior experiments it was dried for 21 days at 23°C and ~ 50% RH. Experiment in tensile mode was made form -100 °C to 100°C at a heating rate 2°C min. Storage modulus E' and loss factor tanδ were recorded.
Microstructure at fracture surface of samples was examined by using Mira/LMU field emission scanning electron microscope (TESCAN a.s.). Dried samples were sputter-coated with gold by using Emitech K550X sputter coating unit (QUORUM TECHNOLOGIES Ltd). The accelerating voltage was 5 - 15 kV.

4. Results

4.1 MS Polymer/Epoxy blends

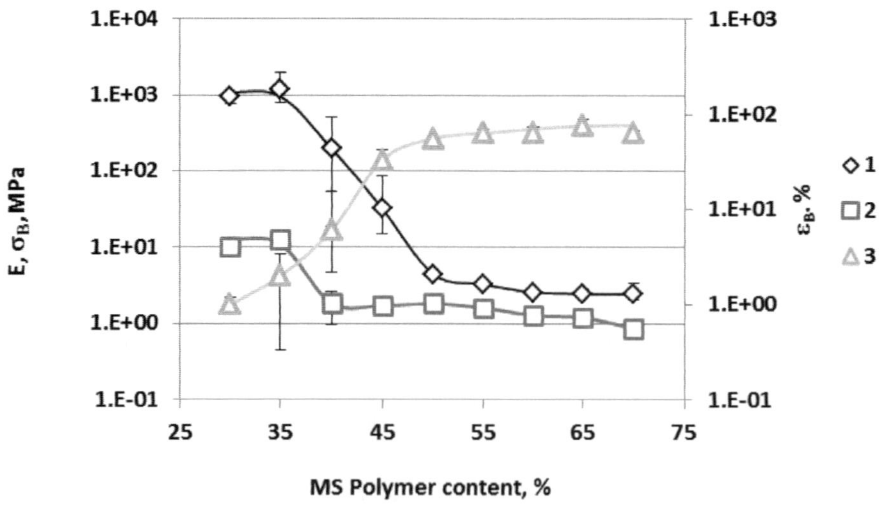

Figure 1. Elastic modulus E (1), tensile strength σ_B (2) and tensile deformation ε_B (3) of MS polymer/epoxy blends with MS polymer/epoxy ratio 30/70 – 70/30.

4.1.1 Mechanical properties. As illustrated in Fig. 1, MS Polymer/epoxy blends possess gradual decrease of elastic modulus E and tensile strength σ_B by increasing MS Polymer ratio from 30 to 70%. At the same time tensile deformation ε_B starts to show considerable values at MS Polymer/epoxy ratio 45/55 and further increases up to ca 60%. The 45/55 ratio is where ε_B overcomes brittleness border of the samples.
Shore A hardness of the discussed MS Polymer/epoxy blends decrease by increasing MS Polymer content presenting almost linear relationship descending from 80 to 38 (Fig. 2).

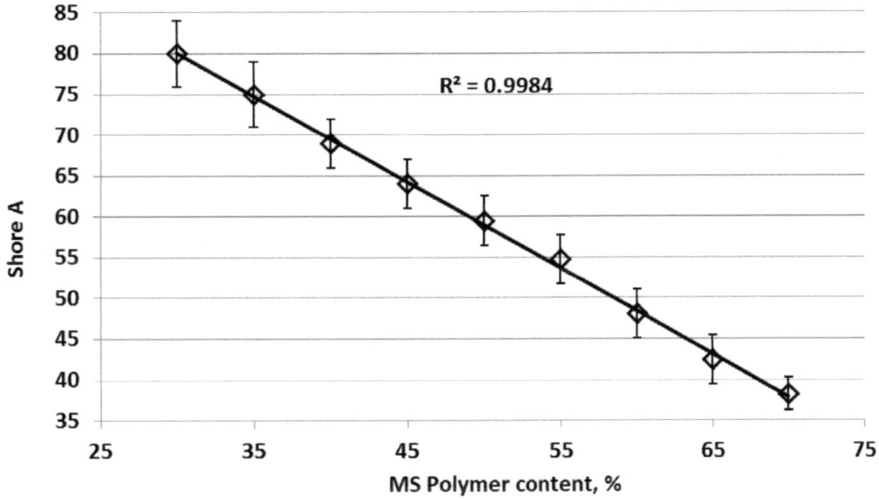

Figure 2. Shore A for MS polymer/epoxy blends with MS polymer/epoxy ratio 30/70 – 70/30.

MS Polymer/epoxy blend with wt.-to-wt. ratio of 60/40 has been chosen for development of adhesive formulation by considering that this composition possess stable and reproducible mechanical properties.

4.2. MS Polymer/Epoxy adhesive formulation

4.2.1. Rheological characteristics. Constant shear rate tests of the formulations of A and B components, as given in Table 3, have been performed. Freshly mixed components A and B reveal thixotropic behaviour as it has been confirmed at constant shear rate test (Fig. 3). At shear rate of 0.2 s^{-1} viscosities of component A and B are 1313 Pa·s and 3176 Pa·s respectively. At shear rate of 10 s^{-1} viscosities of components A and B have been decreased to 75 and 122 Pa·s, respectively, which proves to be suitable for processing of 2 component adhesive. In general, by increasing shear stress to the level common for processing in static extruders or automated mixing equipment, viscosities of formulations of both components approach each other, which is important for ensuring better mixing.

4.2.2. Mechanical properties. During curing of two-component MS Polymer/Epoxy adhesive formulation within the time frame of 28 days gradual increase of tensile strength and decrease of tensile elongation (Fig. 4.) is observed. The greatest changes of ultimate tensile stress-strain characteristics of the investigated adhesive formulation have been observed for curing times up to 14 days. Although afterwards the curing still continues by causing certain increase of σ_B and decrease of ε_B, it can be assumed that adhesive has achieved its final strength after 21 days, as can be judged from approximation of the experimental data by exponential relationships. In total σ_B has been increased by 130% and ε_B has been decreased by no more than 46%.

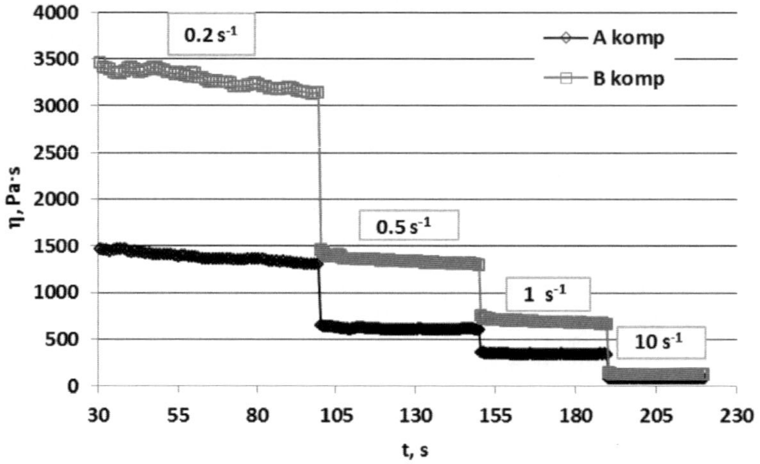

Figure 3. Change of viscosity η at various shear rates for components A and B of MS Polymer/epoxy adhesive formulations.

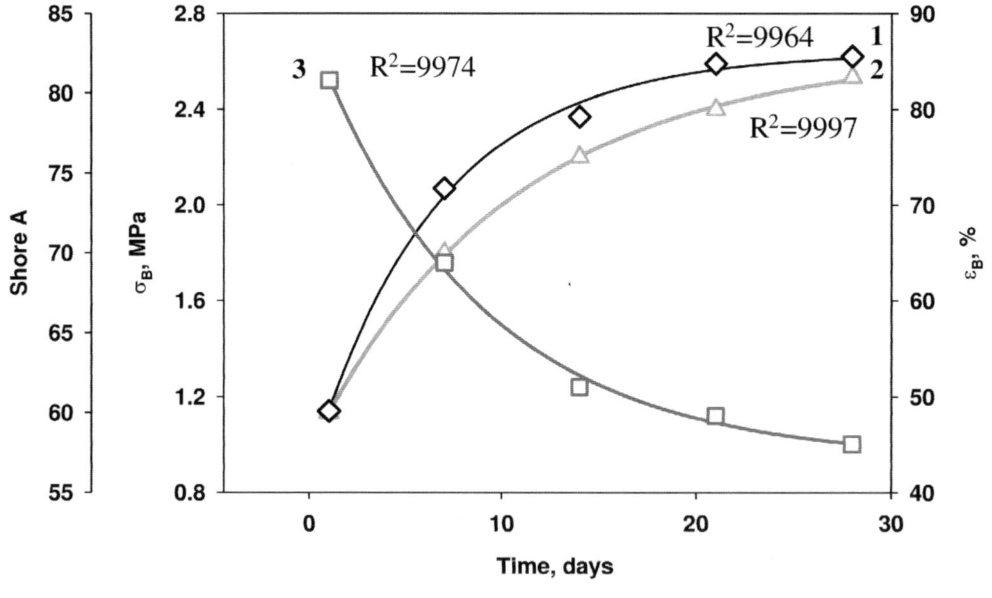

Figure 4. Shore A hardness (1), tensile strength σ_B (2) and tensile elongation ε_B (3) for two-component MS polymer/epoxy adhesive formulation after different curing duration.

Similarly to σ_B Shore A hardness exponentially increases by raising curing time from 1 to 28 days. Already after one day of curing Shore A hardness becomes equal to 60 wich is promising value of initial hardness for adhesive. After 28 days Shore A hardness has been increased up to 81 in average. Increase of Shore A hardness along course of cure, however, is somewhat greater in comparirson to changes in strength of the adhesive.

IOP Conf. Series: Materials Science and Engineering **111** (2016) 012017 doi:10.1088/1757-899X/111/1/012017

4.2.3. DMTA analysis. Tanδ curve of two-component MS Polymer/Epoxy adhesive formulation, shown in Fig. 6, presents two peaks. The first tanδ peak at – 47°C indicates to glass transition temperature (T_g) of MS Polymer and the other peak at 63°C indicates to T_g of DGEBPA epoxy resin. Temperature region from MS Polymer T_g to epoxy T_g could be considered as safe adhesive usage interval where material maintains both required strength and flexibility. As shown in Fig. 6, this temperature interval is between -30°C and 45°C. As expected, below T_g of both components storage modulus E' presents the highest values until T_g of MS Polymer is reached at which modulus decrease by 300 times is experienced. After passing T_g of MS Polymer gradual decrease of E' occurs until reaching T_g of epoxy phase at which E' experiences 12 fold decrease, finally reaching approximately 2 MPa great value.

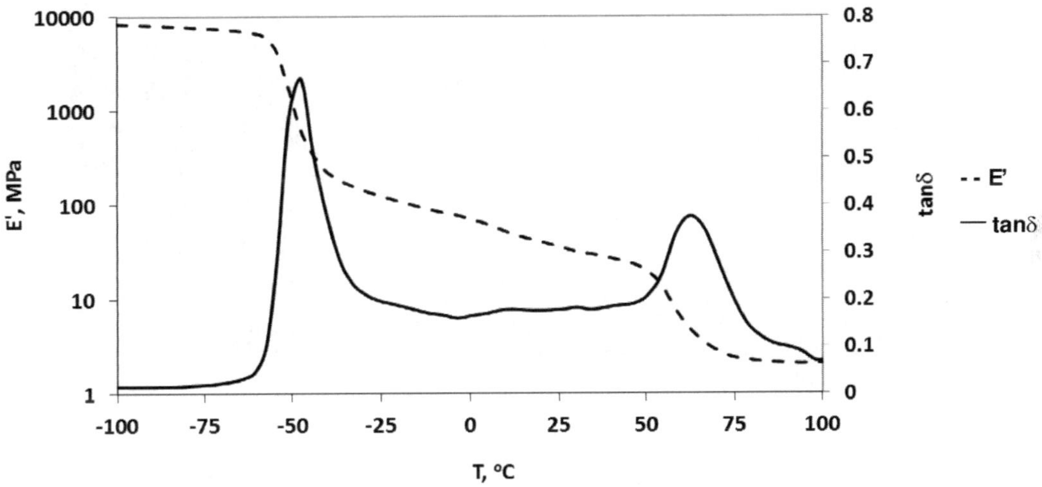

Figure 6. DMA plot of cured MS Polymer/epoxide adhesive

4.2.4. Morphology. SEM micrographs of fracture surfaces of MS Polymer/epoxy 60/40 blend and MS Polymer/epoxy adhesive formulation are shown in Fig. 7. As demonstrated in Fig. 7 spherical epoxy particles are uniformly distributed in the MS Polymer matrix. At higher magnification (Fig. 7. b) it is clearly evident that average size of epoxy particles is less than 10 μm. Although adhesion between MS Polymer and epoxy seems to be good (epoxy particles are smoothly covered with MS polymer without any sign of cracking), some phase separation is possible during fracture as shown in Fig. 7. c. Fracture surface morphology of MS Polymer/epoxy formulation, shown in Fig. 7.d, is more complex due to the presence of broad range of different additives (generally fine and coarse fractions of $CaCO_3$). It is also evident that at the presence of rigid inorganic fillers more brittle fracture is observed.

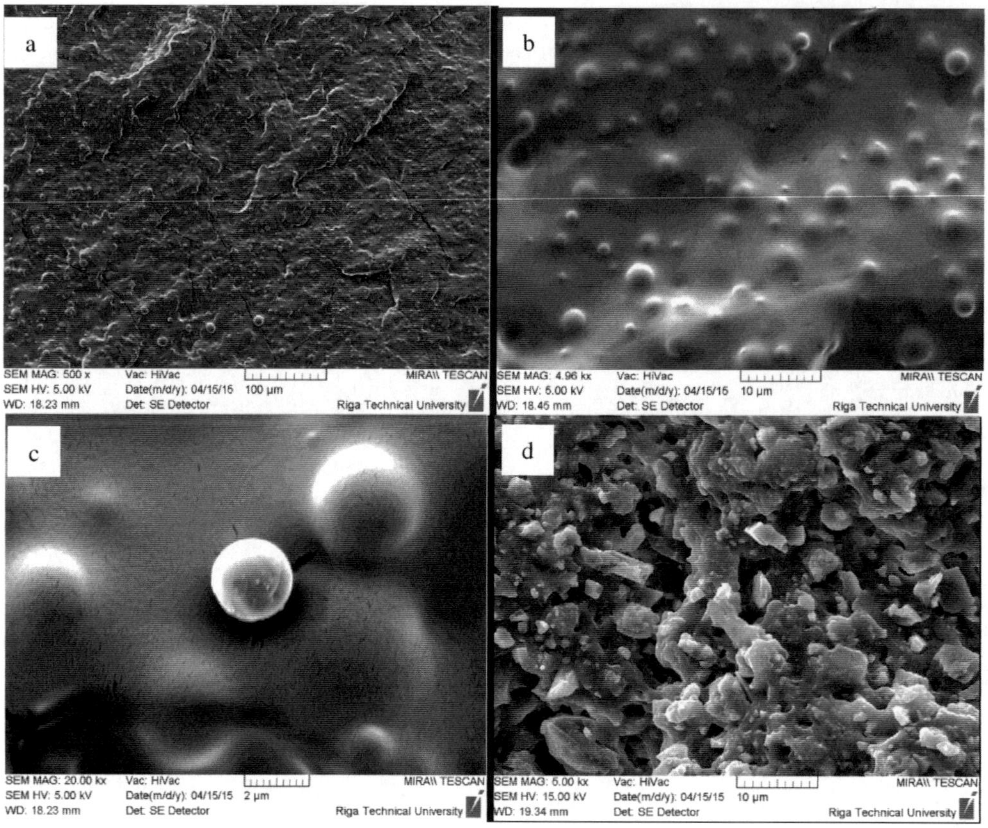

Figure 7. SEM micrographs of MS Polymer/epoxy 60/40 blend at a) 500x, b) 5000x, c) 20 000x magnifications and d) MS Polymer/epoxy formulation at 5000x magnification

5. Conclusions

MS/epoxy blends at wt.-to-wt. ratios from 70/30 to 30/70 are investigated. It is observed that Shore A hardness and strength of the discussed blends decrease, while tensile deformation increase along with raising MS Polymer content.

MS Polymer/epoxy blend at 60/40 ratio has been investigated more in detail to evaluate its applicability for adhesive formulation. The chosen two-component adhesive formulation after 21 days of curing possesses following characteristics: dynamic modulus E' at 23°C – 38 MPa; tensile strength σ_B – 2.59 MPa; tensile deformation ε_B – 48%; Shore A hardness – 79; glass transition temperature T_g of MS Polymer phase - -47°C; glass transition temperature T_g of epoxy phase - 63°C; theoretical temperature interval of usage - -30°C-45°C.

6. Acknowledgements

This work was supported by EU-Project No 2013/0070/2DP/2.1.1.1.0/13/APIA/VIAA/004

7. References

[1] Frigione M E, Mascia L and Acierno D 1995 *Eur. Polym. J.* **31** 1021
[2] Okamatsu T and Ochi M 2002 *Polymer* **43** 721
[3] Thomas R, Yumei D, Yuelong H, Le Y, Moldenaers P, Weimin Y, Czigany T and Thomas S 2008 *Polymer* **49** 278
[4] www.kaneka.be
[5] Devroey D R E, Homma M 2001 *Int. J. Adhes. Adhes.* **21** 275

IOP Conf. Series: Materials Science and Engineering **111** (2016) 012018 doi:10.1088/1757-899X/111/1/012018

Rheological, mechanical and adhesion properties of two component adhesive based on modified silyl terminated polyether polymer and epoxy resin

J Bitenieks[1], R Merijs Meri[1], J Zicans[1], R Berzins[2], J Umbraško[2] and U Rekners[2]
[1]Institute of Polymer Materials, Faculty of Material Science and Applied Chemistry, Riga Technical University, Riga, Latvia
[2]TENACHEM Ltd., TENAX group, Dobele, Latvia

Abstract. 4 types of two component adhesives based on modified silyl-terminated polyether polymer (MS Polymer) and epoxy resin are formulated. Rheological and mechanical properties as well as weathering resistance of the developed adhesives are investigated. It is shown that curing times vary in a wide range depending on the type of MS Polymer. It is also demonstrated that kinetics of cross-linked network formation of the investigated adhesives, depending on the type of MS Polymer, considerably affects mechanical characteristics of the system: Shore A hardness, tensile strength and tensile elongation. Although under the influence of UV weathering mechanical properties of the investigated systems are decreased, the remaining tensile elongation values are not less than 30%. In parallel it is revealed that adhesive formulations, developed in the framework of this research, are most suitable for bonding with stainless steel, partially suitable for bonding with aluminium and less suitable for bonding with polyvinyl chloride.

1. Introduction

Modified silyl-terminated polyether polymer (MS Polymer) as a widespread material in adhesive industry may offer advantages over two-component polyurethane adhesives. MS Polymer has a polyether backbone that carries telechelic methoxysilyl groups, which can crosslink by moisture from the air under ambient temperatures. The polymerisation reaction of MS Polymer is provided by tin catalyst. The elastic behavior of MS Polymer, its inherent good adhesion and stability in combination with epoxy resin has a potential for use in several application areas. Typical epoxy resins usually show an elongation at break of less than 10%. On the other hand their unmatched strength, better heat resistance, reduced tendency to swell when immersed in water, as well as good adhesion to some specialty substrates are properties that are highly appreciated. The combination of epoxy resins with MS Polymer is supposed to deliver a blend that shows superior strength compared to pure MS Polymer based adhesives, while combining all the other positive features from both polymer resin types [1 - 3].

The two-component MS Polymer/epoxy system can be made by incorporating MS Polymer and epoxy hardener in one component and epoxy, MS Polymer catalyst and water in other component [4]. The system is expected to be superb substitute for many (moisture-cure) one-component systems, which can cause problems through slow hardening processes.

Main advantages of such two-component systems are that they can be made for fast curing to eliminate waiting time especially on large work-pieces. Other benefits include high initial strength and cost effectiveness that are important in modern and automated production [5 - 6].

2. Materials

To evaluate the MS Polymer possibilities for two-component adhesive development various types of MS Polymer grades, as shown in Table 1 and Fig. 1, were chosen depending on their physical properties and functionality.

Table 1. Properties of MS Polymer types (manufacturer's information)

MS Polymer type	Structure	Modulus, MPa	Viscosity, Pa·s
SAX 350	DMS	0.25 at 50% elongation	6
SAX 530	TMS	0.27 at 50% elongation	7
S327	DMS	0.16 at 100% elongation	34
SAX 520	TMS	0.33 at 100% elongation	46

a) 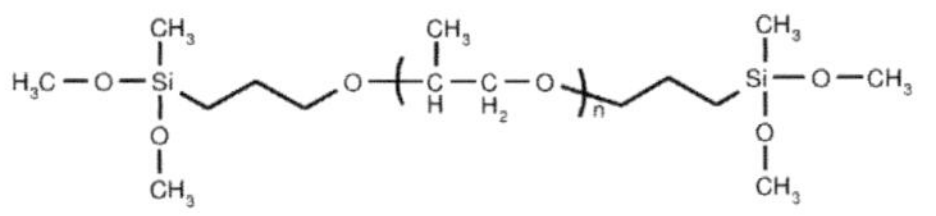

b) 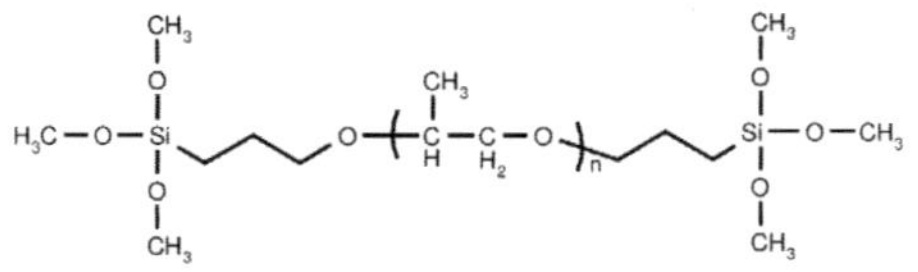

Figure 1. Structural formulas of a) di-functional MS Polymer (DMS) and b) tri-functional MS Polymer (TMS)

General purpose bisphenol A diglycidyl ether (DGEBPA) epoxy resin DER 331 was chosen for the second component of the system. Other ingredients were reinforcing fillers, rheology modifier, silane compatibilizer, antioxidant, catalyst, hardener and UV-stabilizer (Table 2).

Table 2. Materials used for two-component adhesive formulations

Ingredient	Type	Supplier
SAX 350 SAX 530 S327 SAX 520	MS Polymer	Kaneka Belgium
DER 331	Epoxy resin	DOW Chemical
Tib Kat 318	MS Polymer catalyst	TIB Chemicals
Baxxodur EC 130	Epoxy hardener	BASF
AMMO	Aminosilane	Evonik Industries
Omyacarb 1T VA	Coarse filler	Omya AG
Hakuenka CCR-S10	Fine filler	Shiraishi-Omya GmbH
Cabosil TS-720	Thixotropic additive	Cabot Corporation
Irganox 245	Antioxidant	BASF
Tinuvin 328	UV-stabiliser	BASF

3. Experimental

For the development of adhesives (Table 3) total of four different formulations of component A were considered. Component B was formulated on the basis of DER 331 epoxy. Two different formulations of component B were prepared depending on the type of MS Polymer (DMS or TMS) in the component A. For TMS containing systems tin catalyst was added ten times less to obtain reasonable pot life of the system after mixing of components A and B since TMS has three reactive methoxysilyl groups on each end of polyether chain whereas DMS has only two methoxysilyl group on each end of polyether chain.

Table 3. Two-component formulations for MS Polymer/Epoxy adhesive

Formulation for component A			Formulation for component B		
Ingredient	Type	wt, %	Ingredient	Type	wt, %
SAX 350/S327	MS Polymer	30.00	DER 331	Epoxy resin	15.00
Baxxodur EC 130	Epoxy hardener	4.50	Tib Kat 318	Catalyst	1.00
Irganox 245	Antioxidant	0.30	Irganox 245	Antioxidant	0.15
Tinuvin 328	Stabiliser	0.30	Tinuvin 328	Stabiliser	0.15
Omyacarb 1T VA	Coarse filler	19.30	Omyacarb 1T VA	Coarse filler	14.70
Hakuenka CCR-S10	Fine filler	9.50	Hakuenka CCR-S10	Fine filler	2.00
Cabosil TS-720	Thixotropic agent	1.50	H_2O		0.30
AMMO	Compatibilizer	1.30			

Formulation for component A			Formulation for component B		
Ingredient	Type	wt, %	Ingredient	Type	wt, %
SAX 520/SAX530	MS Polymer	30.00	DER 331	Epoxy resin	15.00
Baxxodur EC 130	Epoxy hardener	4.50	Tib Kat 318	Catalyst	0.10
Irganox 245	Antioxidant	0.30	Irganox 245	Antioxidant	0.15
Tinuvin 328	Stabiliser	0.30	Tinuvin 328	Stabiliser	0.15
Omyacarb 1T VA	Coarse filler	19.30	Omyacarb 1T VA	Coarse filler	15.60
Hakuenka CCR-S10	Fine filler	9.50	Hakuenka CCR-S10	Fine filler	2.00
Cabosil TS-720	Thixotropic agent	1.50	H_2O		0.30
AMMO	Compatibilizer	1.30			

Components A and B of MS Polymer/Epoxy adhesive each was mixed with SpeedMixer DAC 150 centrifugal laboratory mixer to obtain homogeneous mixture of the ingredients. Components A and B were then mixed at a ratio of 2:1 by weight. Samples for mechanical characterisation were casted in Teflon moulds and left for curing for determined duration at 23°C ± 2°C and ~ 50% ± 5%RH.

Rheological characteristics were measured using Bohlin CVO 100 rheometer. Instrument was equipped with 20 mm diameter spindle in plate-plate geometry. The gap size between plates was set to 1000 μm. Tests at 23°C were performed in oscillatory shear mode at frequency 1.5 Hz and strain 0.15. The dynamic storage G' and loss moduli G" were recorded as a function of the reaction times.

Tensile stress-strain and adhesive characteristics were determined using Zwick/Roell Z010 universal testing machine. Tensile tests were made according to DIN 53504 at a test speed 200 mm/min. Samples were tested after 14 days of curing. Lap shear tests were made according to EN 1465. Some of the most commonly used materials – stainless steel grade 1.4301, aluminium grade AlMg1 and polyvinylchloride (PVC) grade KömaDur - were chosen as bonded joint substrates. Sample bonding area was 12.5x25 mm with adhesive layer thickness 1 mm. The joint strength was expressed as a maximum force applied to the adhesive area. To observe the development of lap shear strength the tests were made 1, 7, 14, 21 and 28 days after manufacturing.

Shore A hardness was measured according to ISO 868 using Zwick/Roell 3114/5 durometer.

Weathering of adhesives was tested according to ISO 4892-3 using Q-Lab QUV accelerated weathering tester. The test samples were cured for 21 days and then were subjected to periodically repeating cycles of 8 h UV exposure (0.76 $W \cdot m^{-2} \cdot nm^{-1}$ irradiance at 340 nm wavelength) and 4 h condensation. Temperature at UV exposure was 45°C and at condensation step was 40°C. Overall test duration was 672 h. Samples for mechanical testing according to DIN 53504, were taken after 7, 14, 21 and 28 days of weathering.

4. Results

4.1. Rheological characteristics

As seen from the G', G" curves in Fig. 2, in the beginning of time sweep analysis mixtures of the components A and B of the MS Polymer/epoxy system are in liquid state and the values of G' and G" are small. Consequently in the beginning of the experiment viscous behaviour of the systems dominates (G" > G'). Later with progression of curing reaction both moduli increase to the moment where elastic modulus starts to take advantage over viscous modulus. The time where G' crosses G" (G' = G") is then recorded as a gel point (t_{gel}). At t_{gel} stable 3D network is formed however material is not completely cured. After intersection of G' and G", curing starts to develop sharply (G' > G"). Curing behaviour is different for the systems based on TMS and DMS polymers. In comparison to the systems with DMS polymers, systems with TMS polymers exhibit faster increase of G' and G" in the beginning of the experiment, but slower modulus gain after t_{gel} is reached. After t_{gel} curing may proceed up to many hours when G' and G" modules reach plateau. Lower t_{gel} values can be achieved with more reactive TMS MS Polymers like SAX 530 and SAX 520. It is worth mentioning that the use of low-viscous systems with fast curing times could give considerable advantages for development of perspective easy workable adhesives, highly demanded in the market [7].

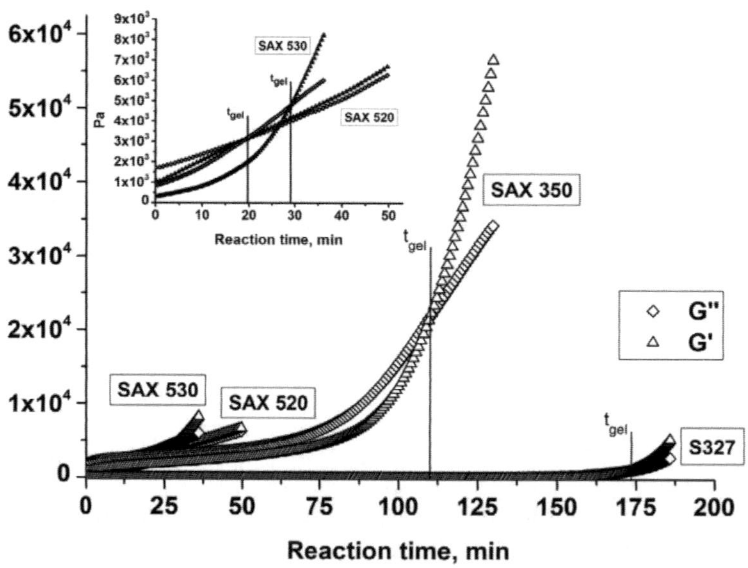

Figure 2. Gel point t_{gel} depending on MS Polymer type.

4.2. Mechanical properties

As already expected, combination of MS Polymer with epoxide resin at the formulations denoted in Table 2 allows obtaining the systems with increased strength and hardness in comparison to neat MS Polymers (10-fold increment in average) [8], thus following the market trends on the developing of high-strength adhesives for high demanding applications [7].

Shore A hardness of the developed MS Polymer/epoxy adhesives is shown in Fig.3. Although all the MS Polymer containing systems show relatively high levels of Shore A hardness already after the 1st day of curing, development of 3D cross-linked network proceeds during the whole observation time, i.e., 28 days, as can be judged from continuously increasing Shore A hardness values. Increase of Shore A hardness of the investigated adhesives can be approximated by exponential relationships. Curing is most intensive during the first 7 days.

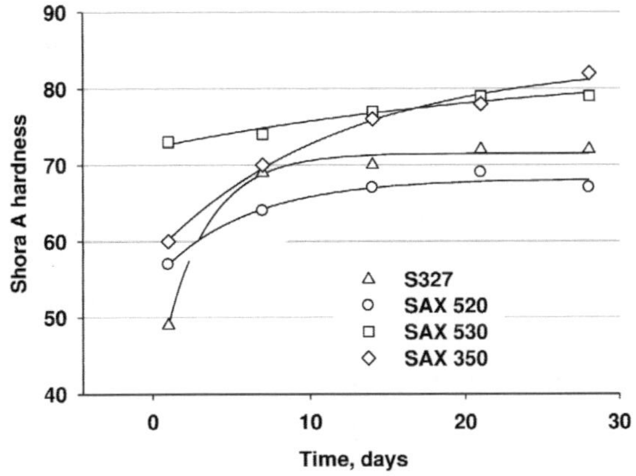

Figure 3. Shore A hardness of MS Polymer/epoxy adhesives.

Tensile stress-strain characteristics of the investigated MS Polymer/epoxy adhesives vary in a similar manner to Shore A hardness. In Fig. 4, tensile stress-strain characteristics of the investigated adhesives are shown after 14 days of curing. The systems containing low viscosity MS Polymers SAX 350 and SAX 530 provide strengths close to 2.5 MPa (namely 2.37 and 2.39 MPa, respectively), while the systems, containing higher viscosity MS Polymer grades, S327 and SAX 520, show ultimate strengths in the range of 2 MPa (namely 1.98 and 1.91MPa respectively). In the same time adhesives with S327 and SAX520 possess higher tensile elongations in comparison to those containing SAX 350 and SAX 530.

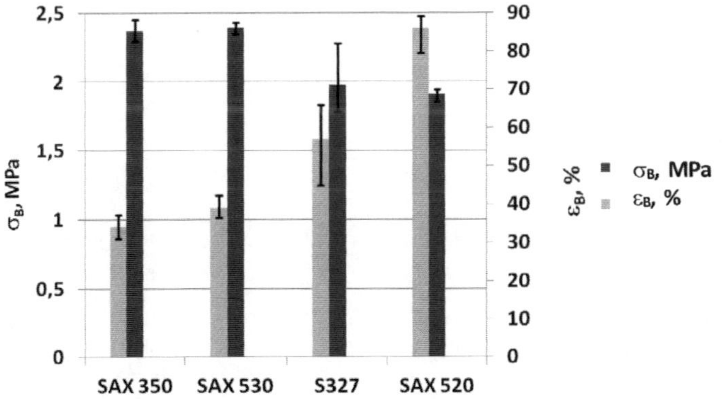

Figure 4. Tensile strength σ_B and tensile elongation ε_B of MS Polymer/epoxy adhesives.

In general Shore A hardness as well as tensile strength and elongation variations denote that main factors determining mechanical characteristics of the developed adhesives are interaction between epoxy resin and MS Polymer during development of the adhesive systems as well as curing kinetics itself.
Lap shear strength experiment was applied as a basis for making some predictions about the reliability of adhesives formulated on MS Polymer. As shown in Fig. 5, increase of adhesive strength to various substrates (stainless steel, aluminium, PVC) is gradually rising along with progression of the cure as well as development of bonding between the adhesive and substrate. Change of adhesive strength of the investigated formulations can be approximated by exponential functions.

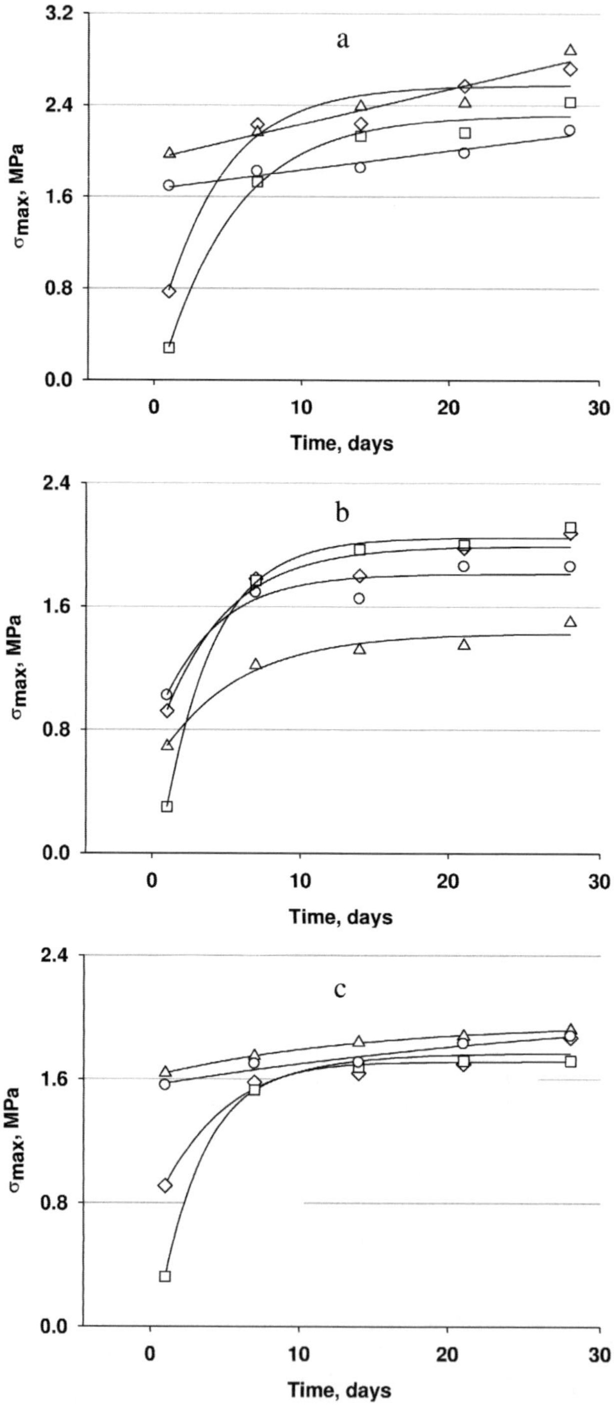

Figure 5. Lap shear strength σ_{max} of MS Polymer/epoxy adhesives to steel (a), aluminium (b) and polyvinyl chloride (c): ◊ - SAX350; □- S327; Δ- SAX530; ○- SAX520.

By considering the importance on the development of fast curing adhesives it is interesting to look on the change of adhesion strength within the bonded system as a function of reaction time. As shown in

Fig. 5, already after the 1st day of curing some of the adhesives show reasonable strength to certain substrates, e. g. formulations with SAX 530 and SAX 520 MS Polymers to stainless steel and PVC substrates. Adhesion of the formulations containing di-functional MS Polymers, especially S327, to these substrates is somewhat smaller. Although after the 1st day of curing adhesion of the investigated adhesives (especially in the case of tri-functional MS-Polymer based systems) to aluminium substrate in general is lower, situation changes after 7 days of curing. Since this adhesive strengths of the investigated formulations (except of the adhesive formulation based on SAX 530) to aluminium and PVC substrates are rather close each to other. Simultaneously adhesion strengths of all the investigated adhesives to steel are the highest; especially this is attributed to the systems containing MS Polymers with lower viscosities. It is also worth mentioning that above 7 days of curing changes of adhesive strengths of the investigated formulations to various substrates are comparatively small. By considering the results obtained, it can be concluded that adhesive formulations, developed in the framework of this research, are most suitable for bonding with stainless steel, partially suitable for bonding with aluminium and less suitable for bonding with PVC. In general the effectiveness of the investigated adhesives to the most promising substrate, i.e., stainless steel, can be arranged in following order: SAX 530 (tri-functional low-viscosity MS-Polymer) > SAX 350 (di-functional low-viscosity MS-Polymer) > S327 (di-functional high-viscosity MS-Polymer) > SAX520 (tri-functional high-viscosity MS-Polymer).

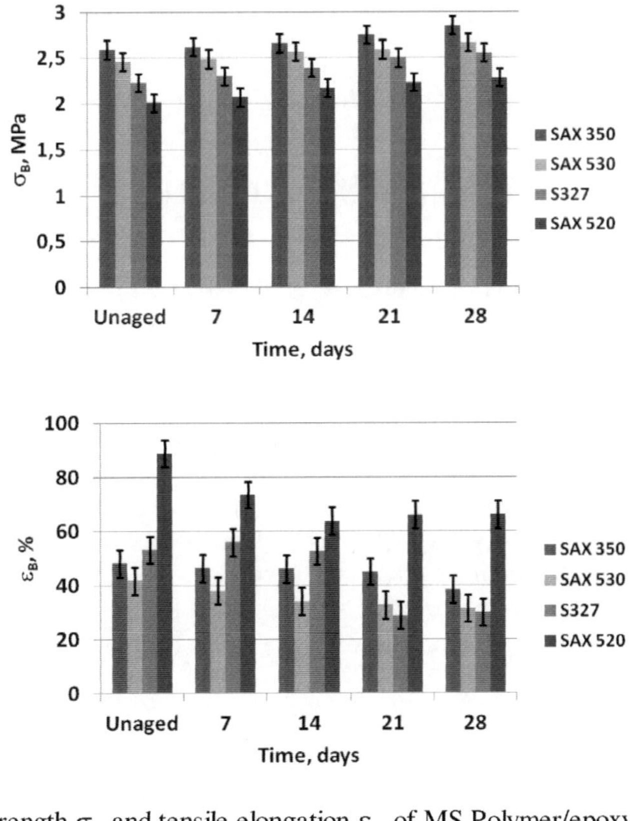

Figure 6. Tensile strength σ_B and tensile elongation ε_B of MS Polymer/epoxy adhesives after UV weathering.

4.3. Weathering resistance

Weathering resistance of the developed adhesives has been evaluated by following the changes of stress-strain characteristics. As shown in Fig. 6, by increasing aging time, tensile strength of all the developed adhesive formulations is slightly increased but respective tensile deformation is decreased in comparison to unaged systems, which is evidently a result of structural impairment of the investigated systems due to the complex behaviour of the detrimental effects of UV irradiation, temperature and moisture. It should be however mentioned that all the investigated adhesive formulations maintained considerable deformations (in the range of 30% at least) even after 28 days of UV exposure. It is also worth mentioning that the composition, containing SAX 530, demonstrating low gel time and possessing the best adhesive properties to steel, showed also one of the highest resistances to UV irradiation (only slightly lagging behind the composition with SAX 350).

5. Conclusions

By investigating rheological and mechanical properties of MS Polymer/epoxy adhesive formulations the following can be concluded:

1) curing times vary in a wide range depending on the type of MS Polymer used;

2) kinetics of cross-linked network formation, depending on the type of MS Polymer, considerably affects mechanical characteristics of the system: Shore A hardness, tensile strength and tensile elongation;

3) although UV weathering detrimentally affects mechanical properties of the investigated systems, the remaining tensile elongation values are not less than 30%;

4) herein defined adhesive formulations exhibit the greatest adhesion to stainless steel, somewhat reduced adhesion to aluminium and the worst adhesion to polyvinylchloride.

6. Acknowledgments

This work was supported by EU-Project No 2013/0070/2DP/2.1.1.1.0/13/APIA/VIAA/004

7. References

[1] Yail J K, Seung W H, Isamu Y and Jae-Yoon K 2014 *Int. J. Adhes. Adhes.* **51** 117

[2] Nečasová B, Kovářová B, Liška P and Šlanhof J 2015 *Procedia Eng.* **108** 410

[3] Acebo C, Fernandez-Francos X, Messori M, Ramis X and Serra A 2014 *Polymer* **55** 5028

[4] Devroey D R E, Homma M 2001 *Int. J. Adhes. Adhes.* **21** 275

[5] Petrie E M 2006 *Epoxy Adhesive Formulations* New York: McGraw-Hill 554

[6] Kohl M, Reichenbach-Klinke R, Steigerwald P, Kern G and Proebster M 2007 *Two-component adhesive/sealant* US 2007/0088110 A1

[7] Peeters L 2013 *adhesion Adhesives & Sealants* **1** 14

[8] Bitenieks J, Merijs Meri R, Zicans J, Berzins R, Umbraško J and Rekners U 2015 *IOP Conf. Series: Materials Science and Engineering* (submitted)

IOP Conf. Series: Materials Science and Engineering **111** (2016) 012019 doi:10.1088/1757-899X/111/1/012019

Impact of non-functionalized and ionic liquid modified carbon nanotubes on mechanical and thermal properties of ethylene-octene copolymer nanocomposites

T Ivanova[1], R Merijs Meri[1], J Zicans[1], A Grigalovica[1], Zh Roja[2], and I Reinholds[2]

[1] Institute of Polymer Materials, Faculty of Material Science and Applied Chemistry, Riga Technical University, Latvia, Paula Valdena Street 3/7, Riga, LV–1048, Latvia
[2] Faculty of Chemistry, University of Latvia, Jelgavas street 1, Riga, LV-1004, Latvia

E-mail: remo.merijs-meri@rtu.lv

Abstract. In this article the development and characterization of composites made from metallocene based ethylene-octene copolymer (EOC) with 38 % octene content, non-modified or modified multi walled carbon nanotubes (MWCNTs), covalently functionalised with long chain hexadecyl moiety imidazolium ionic liquid (IL-f-MWCNTs), is presented. The procedure of MWCNTs functionalization is discussed. In order to obtain a good dispersion of the filler, composites with MWCNTs and IL-f-MWCNTs in the concentration range of 0.5-12 wt.% were made by ultrasonication / thermoplastic mixing method. The results indicated improvement in mechanical properties with increase of the filler content. The methodology of the development of EOC matrix nanocomposites with IL-f-MWCNTs showed advantages of the conductive material creation with high mechanical stiffness compared to direct melt mixing of EOC with non-modified MWCNTs. A notable enhancement of thermal stability was observed in case of both pristine MWCNTs and IL-f-MWCNTs containing EOC nanocomposites, which was attributed to scavenging action of the nanofiller. Modification of EOC with IL-f-MWCNTs showed somewhat increased efficiency in enhancing overall thermal stability of the composites because of better dispersion of the nanofiller due to compatibilizing effect of IL modifier.

[1] remo.merijs-meri@rtu.lv.
[2] zhenija.roja@lu.lv.

1. Introduction

In last decade, the studies of conductive nanocomposite development have broadened due to increased functionality of several carbon nanofillers, such as carbon black, single and multi-walled carbon nanotubes (SWCNTs, MWCNTs), graphene and others, dispersed in thermoplastic matrices [1-3]. In last decade, ionic liquids (IL) have been widely applied in functionalization of carbon nanomaterials in order to improve specific properties such as ionic and electric conductivity, thermal and mechanical performance of carbon nanofillers and their composites with polymer materials in order to develop advanced materials in engineering, electronics and other needs. Recently, ILs have shown benefits on modifying composites with carbon nanofillers: ILs have been used as plasticizers, coupling-agents or as compatibilizers [4-6].

Thermoplastic-polyolefin elastomers, such as ethylene-octene copolymers (EOCs) have potential in several applications due to their low costs, good processability, regular structure and variable properties, depending on octene content in matrix. Recently several authors have shown benefits of incorporating of MWCNT filler in EOC matrix with potential to develop conductive materials, as well as electronic and sensor devices [7-8].

The present paper deals with functionalizing MWCNTs with ionic liquid and comparing the developed IL-f-MWCNTs with pristine MWCNTs as fillers on improvement of mechanical and thermal properties of EOC.

2. Experimental

2.1. Materials and reagents

ENGAGE™ 8200 ethylene-octene copolymer (EOC) with 38 % of octene, density 0.870 g cm^{-3} and melt flow index (MFI) 5.0 dg min^{-1} (at 190 °C/2.16 kg) was supplied by The Dow Chemical Company. MWCNTs of trademark Baytubes® C150P, produced by acetylene chemical vapour deposition (CVD) method (purity ≥ 95wt.%, average outer diameter 13 nm, average inner diameter 4 nm, length > 1 mm, bulk density 1.3-1.5 g cm^{-3}), were supplied by Bayer MaterialScience.

Chemical purity grade (>96%) concentrated hydrochloric acid (37%), nitric acid (70%), and sulphuric acid (97%) were purchased from Penta (Chrudim, Czech Republic). Reagent grade methanol and acetone were both obtained from LabScan (Gliwice, Poland). Tetrahydrofuran (≥99.0%), 1-(3-aminopropyl)imidazole (>97%), 1-bromohexadecane (>97%), sodium bicarbonate, and sodium hexafluorophosphate (99%), all of reagent grade, were purchased from Sigma Aldrich. All reagents were used without purification. High purity water (15 MΩ cm^{-1}), used in MWCNT modification and fabrication of masterbatches, was generated by Millipore Elix 3 (Millipore). Ultrasound bath (Ultrasons 3000837) of 40 Hz, 50 W was purchased from JP Selecta (Poland). Millipore Durapore PVDF membrane filters with pore size 0.1 μ were purchased from Thermo Fisher Scientific.

2.2. Functionalization of MWCNTs by ionic liquid

The modified procedure of MWCNT surface carboxylation, transforming it to methyl ester and covalent grafting of imidazolium ionic liquid derivate to modified MWCNTs was performed on basis of procedure described by Kim et al. [9].

The scheme for MWCNTs functionalization is shown in Figure 1. The procedures for all stages of MWCNTs modification are discussed below.

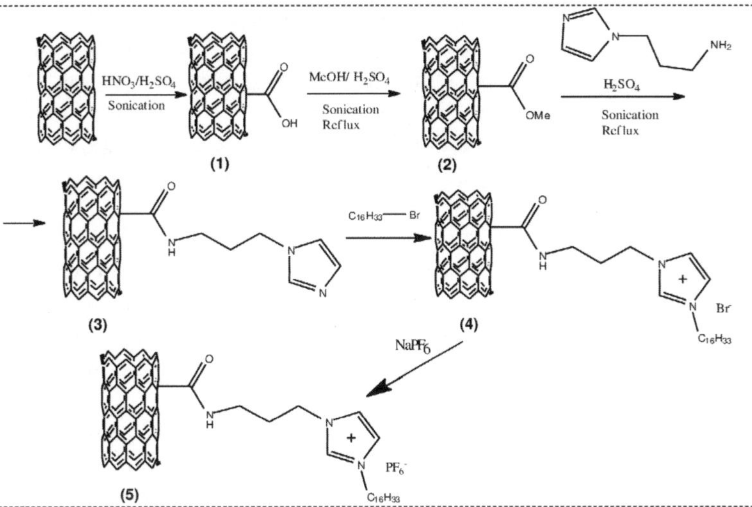

Figure 1. Scheme of MWCNT functionalization by ionic liquid.

Prior to functionalization, the filler was purified by sonicating in 6M hydrochloric acid / 6M sulphuric acid solution mixture for 1 h at room temperature. The obtained purified MWCNTs were subsequently washed with deionized water, filtered through membrane filters and dried in air.

Carboxylated MWCNTs (1) were obtained by functionalization of purified MWCNTs (24 g) with reflux in 1:3 v/v a mixture of sulfuric acid (300 mL) and nitric acid (900 mL). Prior to the process, purified MWCNTs (16 g) were sonicated in the mixture of acids solution for 2 h to open agglomerates. Thereafter, homogenized carbon solution was oxidized under reflux at 100 °C for 6 h to introduce functional carboxyl acid groups. At the end of this period, five-fold dilution was applied to the mixture by deionized water to stop oxidation reaction. The carboxylated MWCNTs were consecutively washed with deionized water until pH of the filtrate was approximately 7. The obtained powder was finally dried in vacuum oven at 60 °C, reaching 15.5 g of final product.

The methyl ester functionalized MWCNTs (2) were obtained by stirring the carboxylated MWCNTs (15.5 g) in methanol with catalytic content of sulphuric acid at reflux under nitrogen atmosphere for 48 h. The reaction mixture was diluted, washed 4 times with deionized water and finally washed with acetone and dried under vacuum overnight at 40°C to give 23.4 g of the product.

Imidazolium functionalized MWCNTs (3) were obtained by stirring mixture of the methyl ester-functionalized MWCNTs (20.0 g) and 480 mL of 1-(3- aminopropyl)imidazole in the presence of a few drops of sulphuric acid at 120 °C for 4 days. Obtained solid product was separated by filtration through a membrane filter, washed carefully with anhydrous tetrahydrofuran, followed by 1 M HCl solution, saturated sodium bicarbonate solution, distilled water until the filtrate pH 7.0, and finally with acetone. The obtained solid matter was dried overnight under vacuum at 50°C to give 19.9 g of Im-f-MWCNTs.

Ionic liquid-functionalized MWCNTs (4) were prepared by stirring the mixture of Im-f-MWCNTs (19.0 g) and 1-bromohexyldecane (190 g) at 80 °C for 24 h under nitrogen atmosphere.

The solids were separated by filtration and washed several times with anhydrous THF to remove the excess bromoalkane. The resulting solids were dried under vacuum at room temperature for 24 h to give 19.0 g of the final product.

In order to obtain more hydrophobic filler, the anion exchange was performed to obtain the final product (5) by the following procedure: 19 g of product (4) was stirred with excess of sodium hexafluorophosphate salt (52 g) in deionised water in room temperature for 24 h. Then the solids were filtered, washed with deionized water for several times and dried for 24 h to obtain the final product of the ionic liquid grafted MWCNT with PF_6^- (IL-f-MWCNTs) as anion (19 g), which was further used for master-batch preparation.

2.3. Composite preparation

Master batches of EOC and MWCNTs (either purified MWCNTs or IL-f-MWCNTs) were prepared by the following ultrasonication procedure: firstly, a solution of 2 wt.% EOC (5 g) in toluene was prepared and ultra-sonicated at 50 °C until obtainment of clear solution. Then calculated amounts of MWCNT or IL-f-MWCNT fillers were added to yield master-batches of composites containing 0, 0.5, 1, 3, 5, 10, and 12 wt.% of fillers in the final compositions. The sonication process was carried out in a thermostatic ultrasound bath for 30 min at 50°C, which allowed to obtain stable dispersions without a tendency of EOC precipitation. EOC based master-batches were obtained by pouring carbon nanofiller containing mixtures in acetone at room temperature under continuous sonication for 1 minute in ultrasound bath. As non-solvent of EOC, acetone led to precipitation of the EOC/MWCNT and EOC/IL-f-MWCNT nanocomposites from toluene dispersions. The products were recovered by vacuum filtration and dried under vacuum at 40 °C. The obtained master-batches were melt mixed with additional amounts of EOC by using two-roll mills for a total time 5 min (speeds of the rolls - 20/25 rpm; temperatures of the rolls - 120/170 °C). Then rectangular plates (length × width ×thickness = 60 mm ×50 mm ×0.8 mm) from the compositions were obtained by compression molding at 120°C.

2.4. Methods of characterization

Infrared spectroscopy was used to confirm the production of ionic liquid grafting to MWCNT surface. The tensile behaviour and thermal properties were characterized to evaluate the influence of ionic liquid grafting to filler surface compared to untreated MWCNTs on the improvement of EOC composite properties as dependence of the filler contents.

2.4.1. Infrared spectroscopy

The Fourier transformation infrared spectrometer (FTIR) Nicolet 6700 (Thermo Fisher Scientifics) was used for confirmation of MWCNTs structure changes during the chemical modification (carboxylation, grafting ionic liquid). The filler powders of non-modified and modified MWCNTs were mixed with KBr. The samples were scanned from 4000 to 400 cm^{-1} at a resolution of 4 cm^{-1}, 24 scans per sample.

2.4.2. Mechanical properties

Tensile tests of all composite specimens were performed at 23±2 °C on a Tinius Olsen H1KS Universal Testing machine equipped with a 1 kN load cell. The compression moulded rectangular plates were cut into conventional dumbbell test specimens with 5 mm working width and 20 mm gauge length. The specimens were pre-stressed to 0.1 MPa before the start of stretching.

The crosshead speed, used for the determination of elastic modulus, was 10 mm min^{-1}, and 200 mm min^{-1} for determination of elongation at break and tensile strength.

2.4.3. Thermogravimetric analysis
Thermogravimetric analysis of obtained composites was carried out by using Mettler Toledo TGA-1/SF equipment at temperature range from 25 to 800 °C at a heating speed of 10 °C min^{-1} under a flow of nitrogen gas. The average sample mass was 10 mg.

3. Results and discussion

3.1. Characterization of MWCNT functionalization
The synthesis scheme of MWCNT modification was successfully realized in order to functionalize the MWCNTs by ionic liquid. The characterization of changes on MWCNT surface was evaluated by the presence of functional group signals in FTIR spectra of pristine MWCNTs and their functionalized derivatives. The average spectra of pristine MWCNTs, the carboxylated MWCNTs and the final product of MWCNTs grafted with ionic liquid are described in Figure 2. The broad band at 3416 cm^{-1} for neat MWCNTs is attributed to stretching vibrations of isolated surface –OH moieties, and absorbed water. In Figure 2c the IR spectrum of IL-f-MWCNTs shows a typical broad doublet of amide -NH moieties at 3400-3500 cm^{-1} range. In all IR spectra the characteristic bands of MWCNTs are shown of C-H covalent and non-covalent vibrations at 3008-2807 cm^{-1} range as well as the bands of C-C backbone at 1586 cm^{-1} and at 1460 cm^{-1} [10]. The bands of C-H vibrations for IL functionalized composites become more intense due to the hexadecane moieties of IL. The intensive bands in the 1750–1550 cm^{-1} range can be assigned to carboxyl groups [10-11]. The appearance of band at 1739 cm^{-1} for carboxylated MWCNTs indicate the oxidation of filler, where the carboxyl group for IL-f-MWCNTs shifts to 1724 cm^{-1} due to the formation of amide group. The overlapped intensive bands at 1621 and 1590 cm^{-1} due to C=N moieties of imidazolium and amide groups of linkers confirm the formation of IL-f-MWNTs [10-11]. The backbone band at 822 cm^{-1} is attributed to imidazolium C-N bounds overlapped with peak of PF_6^- anion vibrations that also confirm the ionic liquid groups in the filler.

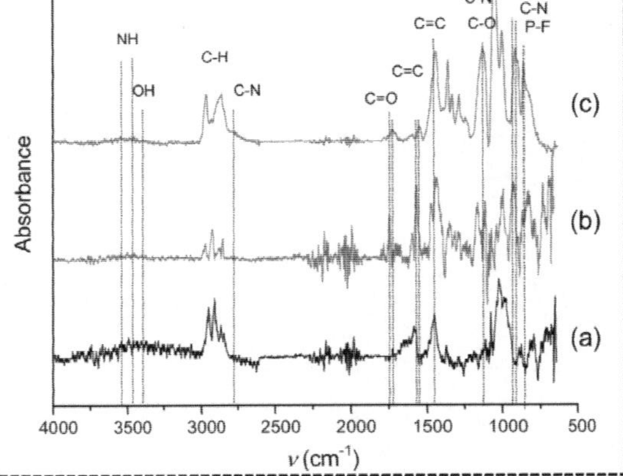

Figure 2. FTIR spectra of neat purified MWCNTs (a), carboxylated MWCNTs (b) and IL-f-MWCNTs (c).

3.2. Stress-strain behavior of composites

The effect of the different MWCNT fillers on the elastic modulus of EOC is plotted in Figure 3. In case of both pristine MWCNTs and IL-f-MWCNTs, the modulus increases steadily with increase of the filler contents until 5 wt.%. The further increase of filler contents from 10 to 12 wt.% show less pronounced growth of modulus, most probably due to fillers agglomeration. As expected, the moduli of EOC composites with ionic liquid grafted MWCNTs are higher compared to pristine MWCNTs containing systems. For example, the modulus increases by 131% to 195%, with increase of pristine MWCNT content up to 3 wt.% and 12 wt.% in comparison to neat EOC, whereas in case of IL-f-MWCNTS the increments reach 152% and 208% at the same filler contents.

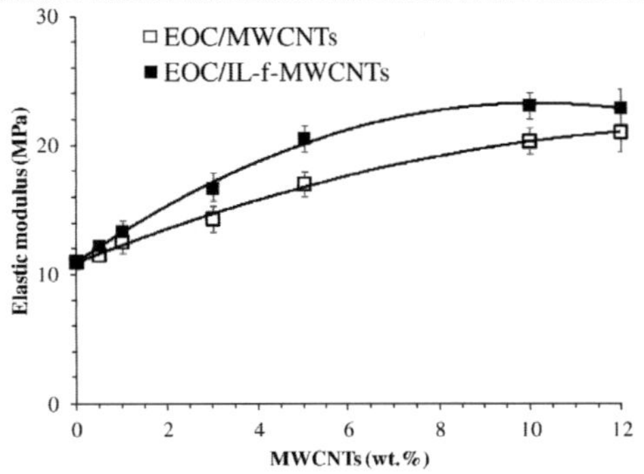

Figure 3. Dependence of the elastic modulus on the contents of the both fillers in EOC composites.

Ultimate stress-strain characteristics of the investigated EOC based compositions are plotted in Figures 4. and 5.

For composites with pristine MWCNTs the dependence of tensile strength, as well as the deformation at break, show a tendency of decrease with increase of the filler content. Insufficient dispersion and possible defect formation in EOC matrix affect the decrease of tensile strength, especially, in case of pristine MWCNTs. However, ultimate stress and strain values of IL-f-MWCNTs containing composites are constantly by 1.2 times higher if compared to the EOC composites with pristine MWCNTs.

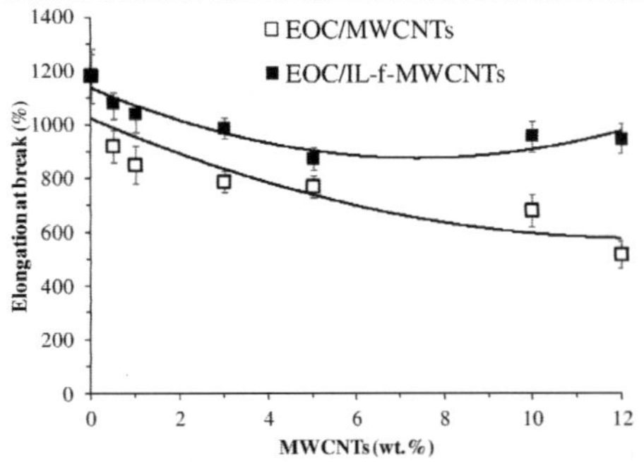

Figure 4. Dependence of the elongation at break on the contents of the both fillers in EOC composites.

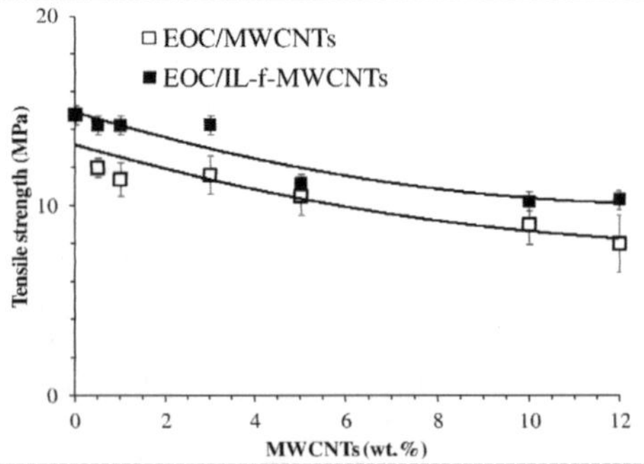

Figure 5. Dependence of the tensile strength on the contents of the both fillers in EOC composites.

IOP Conf. Series: Materials Science and Engineering **111** (2016) 012019 doi:10.1088/1757-899X/111/1/012019

3.3. Thermogravimetric analysis

Degradation of neat EOC and the composites with pristine MWCNTs and ionic liquid functionalized filler were analyzed under inert atmosphere. As MWCNTs possess an antioxidant effect, the increase of thermal stability was predicted with increase of the filler content. It is known that MWCNTs act as radical scavengers within the thermally induced decomposition of EOC chains by breakage of C-H bounds and formation of alkyl and other radicals [12]. In order to investigate the influence of the filler content on EOC composite thermal properties, the values of decomposition temperatures of 10 % mass loss (T_{ons}) as well as the values of decomposition temperatures, at which the greatest mass loss (T_{max}) of the respective decomposition events occur, were obtained from the first derivatives of thermogravimetric curves. The results, compared in Table 1, indicate an increase of T_{ons} (by 21 °C) with increase of filler content from 0 to 3 wt% for EOC compositions with pristine and IL-f-MWCNTs.

Table 1. TGA data of EOC composite decomposition.

ONC (wt.%)	EOC/MWCNT		EOC/IL-f-MWCNT	
	T_{ons} (°C)	T_{max} (°C)*	T_{ons} (°C)	T_{max} (°C)*
0	404	467	404	467
0.5	405	469	404	467/697
1	407	472	408	466/701
3	425	476/620	424	473/736
5	424	481/618	422	472/743
10	423	482/618	407	462/636
12	407	477/618	374	447/642

*value after slash characterizes T_{max} of secondary minor decomposition event

However, the partial agglomeration of MWCNTs at higher weight contents evidently influence a decrease of T_{ons} for both compositions. The main T_{max} value also increases in case of both non-modified MWCNTs and IL-f-MWCNTs containing systems. However, higher values of main T_{max} is constantly observed for the composites with non-modified MWCNTs, reaching maximum 482 °C at filler content 10 wt.%, which is by 14 °C higher temperature than that of neat EOC. T_{max} of secondary peaks are notably higher and emerge at lower filler contents in case of IL-f-MWCNTs in comparison to EOC composites with pristine MWCNTs. That may be attributed to better dispersion of IL-f-MWCNTs due to compatibilizing effect of IL modifier.

4. Conclusions

In the presented studies, commercial MWCNTs were carboxylated and successfully functionalized by grafting surface active imidazolium ionic liquid (IL) with hexadecyl moieties and hydrophobic hexafluorophosphate anion in order to enhance the dispersion of nanofiller in the matrix of ethylene-octene copolymer (EOC). Master-batches of IL-functionalized MWCNTs and EOC were successfully created through the solution casting method followed by melt mixing them into polymer matrix. EOC composites with the content of either MWCNT or MWCNT-IL fillers in the range 0.5-12 wt.% were developed and their mechanical and thermal properties were compared.

The results of analysis showed an increase of mechanical stiffness and thermal stability of EOC composites with both types of nanofillers. IL modified MWCNT-EOC composites due to enhanced interaction of nanofiller particles with polymer matrix showed somewhat improved mechanical and thermal characteristics in comparison to neat MWCNTs containing systems.

5. References

[1] Ma P C, Siddiqui N A, Marom G and Kim J K 2010 Dispersion and functionalization of carbon nanotubes for polymer-based nanocomposites: a review *Compos. A* **41** 1345-1367
[2] Kingston C, Zepp R, Andrady A, Boverhof D, Fehir R, Hawkins D and Wohlleben W 2014 Release characteristics of selected carbon nanotube polymer composites *Carbon* **68** 33-57
[3] Vasileiou A A, Kontopoulou M, Gui H and Docoslis A 2015 Correlation between the length reduction of carbon nanotubes and the electrical percolation threshold of melt compounded polyolefin composites *Appl. Mater. Interfac.* **7** 1624-1631
[4] Peng R, Wang Y, Tang W, Yang Y and Xie X 2013 Progress in imidazolium ionic liquids assisted fabrication of carbon nanotube and graphene polymer composites. *Polym.* **5** 847-872.
[5] Price B K, Hudson J L, and Tour J M 2005 Green chemical functionalization of single-walled carbon nanotubes in ionic liquids, *J Americ. Chem. Soc.*, **127**, 14867-14870
[6] Xu P, Gui H, Hu Y, Bahader A and Ding Y 2014 Dielectric Properties of Polypropylene-Based Nanocomposites with Ionic Liquid-Functionalized Multiwalled Carbon Nanotubes. *J. Electron. Mater.* **43** 2754-2758.
[7] Theravalappil R, Svoboda P, Vilcakova J, Poongavalappil S, Slobodian P and Svobodova D 2014. A comparative study on the electrical, thermal and mechanical properties of ethylene–octene copolymer based composites with carbon fillers. Materials & Design, 60, 458-467.
[8] Petrie K, Docoslis A, Vasic S, Kontopoulou M, Morgan S, and Ye Z 2011 Non-covalent/non-specific functionalization of multi-walled carbon nanotubes with a hyperbranched polyethylene and characterization of their dispersion in a polyolefin matrix *Carbon* **49** 3378-3382.
[9] Kim Y. S, Cha A, Shin J Y, Jeon H J, Shim J H, Lee C and Lee S G 2012 High-density assembly of gold nanoparticles with zwitterionic carbon nanotubes and their electrocatalytic activity in oxygen reduction reaction. Chem. Commun. **48** 8940-8942.
[10] Li Y H, Xu C, Wei B, Zhang X, Zheng M, Wu D, and Ajayan P.M 2002 Self-organized ribbons of aligned carbon nanotubes. Chem. Mater. **14** 483-485.
[11] Maciejewska B M, Jasiurkowska-Delaporte M, Vasylenko A I, Kozioł K K and Jurga S 2014 Experimental and theoretical studies on the mechanism for chemical oxidation of multiwalled carbon nanotubes. *RSC Advances* **4** 28826-28831.
[12] Watts P C P, Fearon P K, Hsu W K, Billingham N C, Kroto H W and Walton D R M 2003 Carbon nanotubes as polymer antioxidants *J. Mater. Chem.* **13** 491-495.

Acknowledgments
The authors acknowledge the Latvian National Research Programme IMIS^2 for the funding of the research.

IOP Conf. Series: Materials Science and Engineering **111** (2016) 012020 doi:10.1088/1757-899X/111/1/012020

Novel alkyd-linseed oil emulsion formulations for wood coatings

E. Sansonetti[1], B. Andersons[2] and I. Andersone[2]

[1] PhD student, Department of Polymer Materials Technology, Faculty of Materials Science and Applied Chemistry, Riga technical University, Latvia

[2] Dr chem., Laboratory of Wood Biodegradation and Protection, Latvian State Institute of Wood Chemistry, Riga, Latvia

E mail: harrysansonetti@hotmail.com

Address. Laboratory of Wood Biodegradation and Protection, Latvian State Institute of Wood Chemistry, 27 Dzerbenes str., Riga LV-1006, Latvia

Abstract. In this experiment an oil-in-water emulsion, suitable as a coating for wood surface protection, was obtained. The main components of this emulsion were linseed oil and an alkyd, mixed at a ratio of 1:2. The experiments were conducted to find a valid combination of solubilizing agents, capable of giving a stable emulsion. Several surfactants, obtained from different producers, were tested. The main factors considered were the nature of the surfactant and its hydrophilic-lipophilic balance (HLB). Our results confirmed that the most suitable surfactants were ethoxylated non-ionic types with a high HLB value, while surfactants based on the gemini technology showed a poor efficiency due to scarce solubility in water, which led to the separation of phases. Once the optimal emulsion composition was obtained, hydrothermally treated and untreated aspen wood samples were coated to evaluate the degree of coating absorption for two different substrates.

Introduction

To protect wood and to improve its appearance and service life, coatings have been used for many centuries. Still today, they occupy a prominent position in the market of wood finishing, and their offer is very wide. Coatings can be distinguished according to the chemical composition of the binder: acrylic, alkyd, polyurethane, nitrocellulose; the destination of use: indoors or outdoors; the nature of solvent, etc. the required properties and, consequently, the formulation of wood coatings commonly depend on the end-use sector: either indoors or outdoors [1]. Coatings can provide wooden materials with the desired aesthetical properties, but are mostly of vital importance in the protection of wood against environmental influences responsible for the deterioration or damage of mechanical or chemical origin. This applies to both interior and exterior applications [2]. When exposed outdoors, wood is subjected to a set of stress factors like light irradiation, thermal radiation, mechanical impact, moisture change and biological attack, causing different weathering effects: photochemical and biochemical degradation, mechanical damages to the surface and colour changes [3]. Coatings for interior use also must withstand different degradation factors like light, climatic variations, mechanical damage like scratches, abrasion, impact and chemical interactions, though conditions are not so harsh like in outdoor exposure: interior wood products require less protection against water and UV radiation than exterior wood products, and finishes usually last for decades [4]. Historically, solvent-borne chemistries dominate wood coatings technologies, and still nowadays, industrial coatings are mainly solvent-borne formulations. However, in the last 20 years, due to the environmental legislation and economic constraints, manufacturers of coatings, ink and adhesives have come under increasing pressure to reduce or eliminate the content of volatile organic compounds (VOC) and hazardous air pollutants (HAP) from formulations, because of the hazard they represent to the environment and the human health [5]. In Europe, this is stated by the Directive 2004/42/EC of the European Parliament and of the Council of 21 April 2004 on the limitation of emissions of volatile organic compounds due to the use of organic solvents in decorative paints and varnishes, amending the already existing Directive 1999/13/EC ("the Paints Directive"). As a result, formulators and users of industrial coatings now face significant challenges as they try to respond to their customers' demands for cost-effective, high performance paints and coatings while meeting increasingly stringent regulations. Actually, there are three leading alternative technologies: water-borne, powder and radiation curable coatings. These relatively new technologies must be improved to keep up with the results of appearance and performance, obtainable by using the solvent-borne technology. Therefore, a continuous improvement of formulations is necessary in the coatings industry in order to satisfy the increasing requirements. In our study, we investigated new formulations for a water-borne coating, using alkyd and linseed oil as binders. Alkyd resins are condensation polymers derived from polyols, a dicarboxylic acid or carboxylic acid anhydride and fatty acids. The name "alkyd" comes from its ingredients; ALcohols and aCIDs, where lately the part CID was replaced with KYD, and the term "alkyd" was created. These resins are polyesters modified with monobasic fatty acids. Alkyds are fundamental binder materials for the manufacture of different types of surface coatings. Alkyd based coatings have good surface coating properties such as good corrosion protection, high gloss, fast dryness, and these resins have good interactions with polar substrates such as wood and steel [6]. Before the introduction of alkyd resins, natural vegetable oils had been used in wood finishing, among them linseed, tung, soybean and safflower oils: they contain unsaturated fatty acids and through a chemical reaction, the oil components crosslink by the action of oxygen, forming a solid film; hence, the definition of drying oils. The most used drying oils are linseed and tung oils. Linseed oil contains unsaturated fatty acids such as linoleic and linolenic acids. it is a relatively inexpensive resource for drying oil, and it provides good protection of wood where the penetration ability of the oil into the wood is superior to other alternatives. However, there are disadvantages such as problems of yellowing, poor long-term stability and rather soft films [7]. This has been the major reason for the replacement of linseed oil by modern synthetic coatings such as alkyds and latex coatings in wood protection; however, linseed oil still plays a role in contributing to the properties of coatings like gloss and elasticity; in wood coatings' formulations, it can be added to alkyd resins. The aim of this study was to obtain an oil-in-water emulsion, suitable as a coating for wood surface protection.

Materials and methods

Composition of the coating

The main components of the emulsion we tried to obtain were distilled water as a solvent, binders and surfactants. Only water was used as a dispersing agent, without a cosolvent, that is usually present also in water-borne coatings. The binders used in this study were linseed oil and alkyd resin they were provided by the Riga Varnish and Paint Factory. Alkyd was prepared heating 55% of tall oil fatty acids, 25% of phthalic anhydride and 20% of pentaerythritol; as a result, a medium length alkyd was obtained. The content of alkyd and linseed oil in the emulsion was about 20%, and their ratio was 2:1, respectively. Samples of 9 surfactants were provided by the Air Products Company: they were selected according to their chemico-physical properties and as recommended by the producer for use in water-borne coating formulations. Their generic characteristics are listed in Table 1. The concentration of the surfactant varied between 1.5% and 2.5%. The Carbowet series surfactants are linear ethoxylated alcohols, with a chain length varying from 9-11 (Carbowet 109, 125, 138) to 12-15 atoms of carbon (Carbowet 144), the Dynol and Surfynol series are acetylenic based surfactants, all containing a triple bond: Dynol 800 and 810 are the same ethoxylated 2,5,8,11-tetramethyl-6-dodecyn-5,8 diol, but at 2 different concentrations, Surfynol 104A is a solution of 50% 2,4,7,9-tetramethyl-5-decyne-4,7-diol in 2-ethylexanol, Surfynol 440 is ethoxylated 2,4,7,9-tetramethyl-5-decyne-4,7-diol, and Surfynol 2502 is ethylene oxide-propylene oxide copolymer diether with 2,4,7,9-tetramethyl-5-decyne-4,7-diol (all surfactants of the Surfynol series are derived from 2,4,7,9-tetramethyl-5-decyne-4,7-diol). The concentration of the surfactants was between 1% and 2%. Other additives were included in the formulation: cobalt drier to promote the oxidative reaction of binders, an anti-skinning agent to complex cobalt drier, thus avoiding top film formation during storage, and an antifoam agent to prevent foam formation during the emulsification process.

Table 1.

Surfactants used and some of their characteristics

Name of surfactant	Type-concentration (%)	HLB	Aspect
Carbowet 109	Nonionic, 60-100	13.1	Liquid
Carbowet 125	Nonionic, 100	12.5	Liquid
Carbowet 138	Nonionic, 100	13.9	Liquid
Carbowet 144	Nonionic, 100	14.4	Solid wax
Dynol 800	Nonionic, gemini, 70	not available	Liquid
Dynol 810	Nonionic, gemini, 80	not available	Liquid
Surfynol 104A	Nonionic, gemini, 50	4	Liquid
Surfynol 440	Nonionic, gemini, 100	8	Liquid
Surfynol 2502	Nonionic, gemini, 100	8	Liquid

Preparation and testing

To facilitate the emulsification, a high shear is necessary, since water and the binders are immiscible. The components were mixed using an ultrasonic homogenizer MRC Sonic-650, with a power of 650 W, set at 70% of total power. Total mixing time was 10 to 20 min; it consisted of repeated sessions of mixing and pauses – 9 s of mixing, followed by 1 s of a pause, respectively. The homogenizer measured the temperature of the sample and to avoid the heating of the emulsion; a bath of ice and water was necessary to keep the temperature low. To verify the drying rate and measure the solid content, 2 g of coating was put in an aluminium plate and let it dry at room temperature, measuring the weight after 24 and 48 h. Coating tests were conducted on samples of aspen (*Populus tremula*) measuring 150x70x15 mm, to verify the applicability and penetration of the coating. Three layers of coatings were applied with a brush, allowing each layer to dry for 24 h. To verify the response of different substrates, coatings were applied on hydrothermally treated and untreated samples. The hydrothermal treatment has influence on the chemico-physical properties of the wood, in particular it decreases its equilibrium moisture content (EMC) and its density: after conditioning hydrothermally treated and untreated samples at a moisture of 65%, the EMC is 4.6% and 10.7% respectively for treated and untreated aspen wood samples. Also a decrease of density was observed: untreated aspen has an average density of 0.524 g/cm^3, after the hydrothermal treatment density decreases to 0.445 g/cm^3, thus becoming lighter and more porous; this is due to the loss of mass after thermal degradation of wood components. The difference in density can affect the penetration of coating. Aspen wood was hydrothermally treated at 170°C for 100 min at high pressure in water vapour medium.

Results and discussion

Emulsification of alkyd and linseed oil

All water-borne coatings need surfactants in order to reduce the free energy of the various interfaces of the system, thus providing kinetic stability to the formulation. Surfactants cover a wide range of functions in coatings – they allow the emulsification of binder in the solvent, facilitate dispersion of pigments, they are needed to improve wetting on low energy substrates, to control foaming during the application and processing, and to prevent film defects, caused by surface tension gradients [8]. The first step was the choice of the right surfactants, among the available ones, to emulsify the binders initially without other additives. Our experiments confirmed that the most suitable surfactants are those of the Carbowet series, ethoxylated alcohols: they are nonionic, with a high HLB, are easily and totally soluble in water, and form stable oil-in-water emulsions with alkyd and linseed oil. Emulsification of alkyd and linseed oil with other surfactants, Dynol and Surfynol, was not successful, due to the separation of phases that rapidly occurred after mixing. Though gemini surfactants have a higher surface activity, owing to their structure, containing two hydrophobic tails and two polar head groups, in this case, they are not effective, due to their scarce solubility in water; this is also confirmed by the lower values of HLB.

Additives and stability of coating

Although used in very small quantities, additives can have a huge impact on coatings' performance and application properties. The most difficult task is to find how to use them in the right order, at the right time, and at the right levels; additives can help improve the appearance and durability of a coating, the flow of paints and the efficiency of the manufacturing process, and other properties. One of the most important groups of additives is that of catalytically active additives, which include paint drying agents; they are added to air-drying coating formulations to catalyse the oxidative cross-linking reactions. In our study, cobalt, calcium and zirconium driers were tested singularly for each emulsion, and the only efficient drier was cobalt salt. Cobalt is a highly effective drier and a strong oxidant. It is usually added at 0.05 to 0.4%, depending on the binders' concentration. At this stage of the experiments, once the suitable surfactants were found, several coatings were prepared to investigate the necessary concentration of cobalt, their drying rate at room temperature, and at which stage cobalt should be added to the coating. The results of drying rate are listed in Table 2. In all cases, coatings dried, and the formation of a film was observed. For EML01 A, EML02 B, cobalt was added at the end, after the emulsification of binders, and then sonicated again to disperse it; the concentration was 0.20%. With this procedure, the cobalt dispersed in the coating, although a longer time of sonication was required, but the emulsion was not stable, and eventual separation of phases occurred in the next days, with oxidation of binders. For the coatings EML03 C, EML04 D, EML05 E and EML06 F, the sequence was changed: cobalt was added to linseed oil and alkyd, before emulsification with water and surfactants. These coatings were more homogeneous; the dispersion was better and more stable.

Table 2.

Measurement of mass changes during the coating's drying at room temperature

EMULSION	**Co (%)**	**Binder (%)**	**m (g)**	**m after 24h (g)**	**m after 48h (g)**
EML01 A	0.20	22.7	2.00	0.51	0.50
EML02 B	0.20	22.7	2.00	0.62	0.54
EML03 C	0.10	17.3	2.04	0.48	0.41
EML04 D	0.17	21.3	2.00	0.41	0.38
EML05 E	0.11	20.9	2.00	0.48	0.40
EML06 F	0.13	21.0	2.00	0.39	0.38

To prevent the formation of a film on top of the coating during storage, an anti-skinning agent was added to the formulation, specifically it is methyl ethyl ketoxime, which by binding of the drying agent, inhibits the oxidative crosslinking reactions of resin and drying oil before its utilization. During the emulsification, the formation of foam at the surface was observed with ethoxylated alcohols; this is an undesired property in the preparation process, and to avoid it, a silicone based antifoam agent was used. It turned out to be very efficient at very low concentrations. This antifoam agent can be added at any stage of the process.

Evaluation of coating penetration

The application of coatings on hydrothermally treated and untreated aspen wood samples resulted in different behaviours. The type of the substrate influenced the wetting properties of the coating and the absorption. A deeper penetration of the coating was observed for hydrothermally treated wood than for untreated wood. For hydrothermally treated wood, the coating penetrated from 5 to 15 mm, whilst for the untreated wood, it penetrated from 1 to 4 mm. This can be explained by the changes in the chemico-physical properties that occur after hydrothermal treatment: wood becomes more hydrophobic, has lower moisture content and its density decreases, making it more porous. This facilitates the absorption of the coating under the surface. Ulterior investigations are necessary to examine the distribution of binders at the wood-coating interface.

Conclusion

Our experiments confirmed that ethoxylated alcohol surfactants are effective dispersing agents in water-borne coatings. They provide a stable emulsion, dispersing alkyd and linseed oil without the disruption of droplets or coalescence. The use of ultrasonication to homogenize alkyd and linseed oil in water led to the formation of a white emulsion. In our experiments, only water was used as a solvent, avoiding the utilization of an organic cosolvent, which is usually present also in water-borne coatings. Cobalt proved to be a highly efficient drier, also at a low concentration, but there are some problems related to the dispersion and stability of the coating, and the sequence of the mixing order, eventually calcium and zirconium can be used together with cobalt drier, but not alone, since they showed a poor efficiency. Additional additives and a combination of surfactants should be further investigated to improve the coating stability. Due to chemico-physical changes of wood, in hydrothermally treated wood absorption and penetration of the coating is deeper than in untreated wood.

References

[1]Scrinzi E et al 2011 Prog. Org. Coat. **72** (1) 81
[2]De Meijer M 2001 Prog. Org. Coat. **43** (4) 217
[3]Chemistry of Weathering and Protection, Ed. by Rowell, R.M. The chemistry of solid wood 1984. Washington DC. American Chemical Society. pp. 401–451
[4]Finishing of wood, in: Wood handbook–Wood as an engineering material 1999. Gen. Tech. Rep. FPL–GTR–113. Madison, WI, U.S.A.: USDA Forest Products Laboratory ch. 15 pp.1–37
[5]Chang, C-W and Lu K-T 2013 Prog. Org. Coat. **76** (7-8) 1024
[6] Ilhan K et al 2014 Prog. Org. Coat. **77** (5) 949
[7]Stenberg C et al 2005 Ind. Crops Prod. **21** (2) 263
[8] Hellgren A-C et al 1999 Prog. Org. Coat. **35** (1-4) 79

IOP Conf. Series: Materials Science and Engineering **111** (2016) 012021 doi:10.1088/1757-899X/111/1/012021

Synthesis of Polymerizable Ionic Liquid Monomer and Its Characterization

Jagannath Sardar[1], Uno Mäeorg[2], Illia Krasnou[1], Vikram Baddam[1], Viktoria Gudkova[1], Andres Krumme[1]*, Natalja Savest[1], Elvira Tarasova[1] and Mihkel Viirsalu[1]

[1] Tallinn University of Technology, Department of Polymer Materials, 19086 Tallinn, Estonia.
[2] University of Tartu, Institute of Chemistry, 50411 Tartu, Estonia.

*Any correspondence to be made: Andres.Krumme@ttu.ee, Phone: +372 6202907

Abstract:
The polymerizable monomer of 1-(6-hydroxyhexyl)-3-methylimiadazolium bromide ionic liquid (IL) was synthesized. The thin layer chromatography tests were conducted at each step to verify the reactions. It was observed that the ionic liquid compound formed after 48 hours though confirmed the presence of starting materials. A high viscous transparent liquid was extracted and believed to be formed of 99% yielding of ionic liquid. Apart from that, the synthesis of ionic liquid was also performed using microwave treatment and heating to stimulate the reaction. Surprisingly, it was observed that the reaction was taking place within a fraction of second when the synthesis process was conducted with microwave treatment. In the second step, esterification IL with methacryloyl chloride was done to obtain monomer of polymerizable ionic liquid (m-PIL). The final product was a high viscous yellowish liquid.

Keywords: Synthesis, polymerization, ionic liquid, polyionic liquid

1. Introduction

The synthesis of polymerizable ionic liquids (PILs) is a challenging objective and has drawn an intense attraction for its uniqueness such as mechanical durability, excellent ionic conductivity and low glass transition temperature (Tg) which make suitable to employ in electrospinning process and other emerging technology [1-2]. Over the fifteen years, researches going on to polymerize the ionic liquids in various methods. It is observed that the ionic conductivity drops repeatedly after polymerisation due to decrease of mobility of ions in the medium [1]. Owing to enhance the electrical conductivity, preliminary IL can be blended with PILs monomer [3]. The PILs can be classified into several events by their polymerization manner such as polyanion and polycation type ILs, polyzwitterion type IL, copolymer type IL, cross-linked PILs etc. PILs can be synthesized via two basic strategies direct polymerization of IL monomers and chemical modification of existing polymers. A polymerizable unit can be located in an ILs monomer conjugated either with cation or with anion, solely depending on the architecture of desired polymeric constituents [4]. These processes can be employed by several methods such as free radical polymerization to homopolymer or cross-linkable ILs, copolymerization, in-situ polymerization/post-processing polymerization and other polymerization methods [4-6].

We made synthesis of imidazolium-bromide based PIL, introducing an effective functional group of the methacylate into synthesized IL backbone capable for next polymerization [2]. The higher alkyl chain lengths between imidazolium cation and polymerizable entity the higher the spacer length which enhance electrical conductivity of the polymerized material [2, 7]. The main objective of the work has been specified by two steps: (a) synthesis of IL using 1-methyl imidazole by alkylation with 6-bromo-1-hexanol and (b) synthesis of PIL monomers using methacryloyl chloride followed by esterification of the alkylimidazole ionic liquid of the previous step.

IOP Conf. Series: Materials Science and Engineering **111** (2016) 012021 doi:10.1088/1757-899X/111/1/012021

2. Materials and Methods

2.1. Synthesis of IL (1-(6-hydroxyhexyl)-3-methylimadazoium)

In accordance to the initial step to synthesis ionic liquids, 1-methylimidazole (Sigma-Aldrich) and 6-bromo-1-hexanol (Aldrich) were used as received. The detailed procedure formulated to synthesis the ILs can be found in the following statements:

Initially, 6-bromo-1-hexanol was taken in a round bottom flask. The weight of the chemical was noted as 4.79 g equivalent to 0.0264 moles, calculated according to molecular weight (Mw: 181.07 g/mole). To keep the molar ratio of 1:1, methylimidazole was measured as 2.166 g equivalent to 0.0264 moles (Mw: 82.10 g/mole). Methylimidazole was added to bromohexanol dropwise. The mixture flask was kept at room temperature filled with Argon gas and stirred mechanically using magnetic stirrer in order to synthesis the alkylated IL. This step takes 48 hours.
To control the reaction mixture the thin layer chromatography was conducted time after time in a predefined time span. The chromatographic tests were also done for the starting materials.
The possible reaction between bromohexanol and imidazole has shown in the Figure 1 the reaction yielded transparent viscus solution.

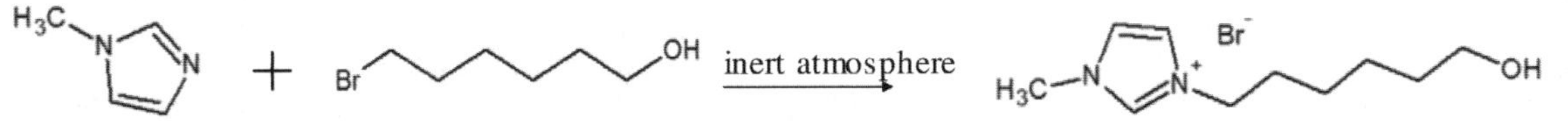

1-methylimidazole 6-bromo-1-hexanol 1-(6-hydroxyhexyl)-3-methylimadazolium bromide

Figure 1: Synthesis of alkylated IL: Chemical reaction and yielded transparent compound

In the next step the IL compound was kept in vacuum for 12 hours at 1.1 mbar pressure in order to remove the unwanted substrates. The final IL was washed out thoroughly with toluene for three times in order to remove the starting materials (imidazole and bromohexanol) and hence vacuumed at the same pressure for another 12 hours. Final yield of the product after vacuumed was measured approximately 6.46 g.

2.2. Synthesis of PIL (1-[(6-methacryloyloxy) hexyl]-3-methylimidazolium bromide)

All next described processes were done under Argon atmosphere. Synthesized 1-(6-hydroxyhexyl)-3-methylimadazoium bromide was taken into the round bottomed flask (6.41g) and suspended into 37.66 ml of acetonitrile. In order to stabilize and prevent localized oxidation 1.25 mg of BHT (2,6-di-tert-butyl-p-cresol) was added into the mixture. The mixture was set on ice bath to prevent uncontrolled exothermal reaction during mixing of components.
Methacryloyl chloride (5.11 g) was taken in another round bottom flask and 5 ml of acetonitrile was added. This solution was added dropwise into the reaction flask with IL. After mixing of components the reaction mixture was leaved to warm up to room temperature and stirred for 48 hr.

Figure 2. Synthesis of PIL: chemical reaction and yielded compound (1-(6-hydroxyhexyl)-3-

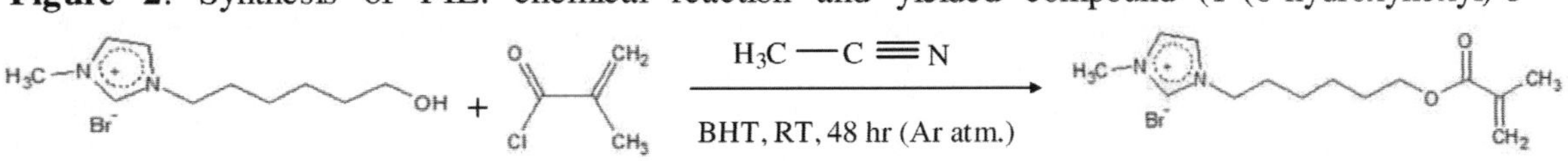

methacryloyl chloride, 1-[(6-methacryloyloxy) hexyl]-3methylimadazoium bromide and methylimidazolium bromide).

After 48 h the solution was dried out using condenser reflux in order to remove unreacted chemicals and gases. In order to remove excess of methacryloyl chloride, BHT and acetonitrile, product was washed for three times with diethyl ether (3 × 10 ml) and kept 12 hours at vacuum. The yielded m-PIL was extracted as high viscous yellowish liquid. The probable reaction was shown in the Figure 2.

3. Thin Layer Chromatography

Thin layer chromatography (TLC) is an advanced technique to identify the chemicals ad compounds. Nevertheless, this technique gives bold information about the nature of the substrates as well as highlights the stability of the chemical structure. It also separate non-volatile mixture and determine the purity of the chemicals [8-11]. This technique boosts the idea in order to determine the polarity of the applied substrates [12].
In the present work, the TLC analyses were carried out on aluminium-backed 65 × 25 × 0.2 mm plate of improved resolution silica sheets (Alugram®Sil G/UV254, Macherey-Nagel). Development was performed in a developer glass chamber with glass lid. The chamber dimensions of 70 mm tall and 40 mm inner diameter, containing only a minimal volume of the eluent solvent.

Initially the substance was dissolved with suitable solvent in 2 ml Eppendorf tube. For this study, specifically, methanol was used to dissolve IL compound and ethyl acetate for starting chemicals, imidazole and bromohexanol. The detailed of suitable dissolved solvent and eluent solvents have been listed in the Table 1.

Table 1. Relevant solvents for dissolution and eluent solvents for TLC

Compound	Solvent used for dissolve	Solvent taken in TLC chamber
1-methylimidazole	ethyl acetate	90% ethyl acetate + 10% methanol
6-bromo-1-hexanol	ethyl acetate	90% ethyl acetate + 10% methanol
1-(6-hydroxyhexyl)-3methylimiadazolium bromide	methanol	90% ethyl acetate + 10% methanol
AIBN (azobisisobutyronitrile)	methanol	85% hexane + 15% methanol
2-hydroxyl-2-methyl propiophenone (Mw: 164.20 g/mole)	methanol	85% hexane + 15% methanol

In the developer chamber, the eluent solvent 90% ethyl acetate + 10% methanol was taken for 1methylimidazole, 6-bromo-1-hexanol and 1-(6-hydroxyhexyl)-3-methylimiadazolium bromide (IL), whereas, 85% Hexane + 15% methanol was taken as solvent for AIBN (Azobisisobutyronitrile) 2-hydroxyl-2-methyl propiophenone.
At the first step of TLC, the compound or the starting substances of 2-4 μL were applied using capillary tube to the TLC plate at marked starting line. The TLC plate was placed vertically with slide inclination in the developer chamber in such a way so that the starting line should not be immersed into eluent solvent. The solvent, thus, allowed to elute through Silica gel contained TLC.
After a certain time, the TLC plate was taken out from the chamber and marked the finishing line of the eluent and examine under fluorescent light. If the spot unidentified under fluorescent light, the TLC plate was plunged into potassium di-cromate ($KMnO4$) or Phosphomolybdenum acid solution to identify the resultant spots.

4. Results and discussion

In the preliminary experiment, imidazole and bromoxehanol were tested with TLC. The mixed solvent contained 90% of Ethyl acetate and 10% of Methanol. Further, the spot was contrasted by exposure in $KMnO4$ solution. The samples of IL were also tested by TLC to identify the starting materials residues. The presence of starting materials in final product proved that the reaction yielding less than 100%. The starting materials removed from the compound after three times wash with toluene.

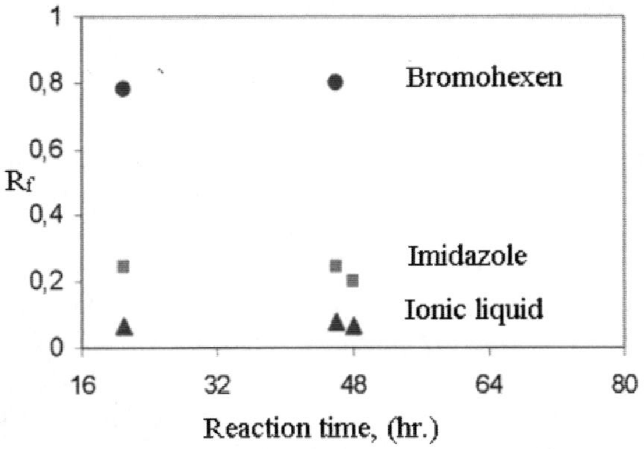

Figure 2. Variation of Rf of starting materials and the synthesized ionic liquid.

In order to predict the polarity behaviour of the chemicals, it was observed that the bromohexanol has the less polar nature as compared to imidazole and the reaction compound i.e. IL has the highest polarity than imidazole as well as bromohexanol. Similarly, the initiator for polymerization, hydroxy propeophenon and AIBN were also used to predict polarity. One can notice that the hydroxy propeophenon has high polarity as compared to AIBN.

From the TLC results the retention factor (Rf) was calculated and plotted in the Figure 2. From the figure, it was observed that the bromohexanol is less polar, as compared to imidazole, whereas ionic liquid shows the highest polarity.

The above mentioned process, in order to synthesize of ILs was also conducted using microwave treatment to reduce the duration of reaction. Eventually, it was observed that the reaction was taking place a fraction of second when microwave was used. The yield compound was extracted as light brown viscous liquid which could evidence decomposition of reaction substances. The TLC tests were also conducted in the similar manner and the results are demonstrated that the starting chemicals are present in the yield compound of ILs in a significant amount. It is to be believed that the starting materials can be removed by washing of the reaction compound using toluene.

The heat induced synthesis of IL was done temperature at 80 °C. In this process the time was predefined for 24 hours. The mixure of the starting chemicals, imidazole and bromoethanol were taken in a round bottom flask and kept in an oil bath in order to maintain exact temperature. The IL compound was turned to a solid salt when it contacted the air, presuming the oxidation of the compound.

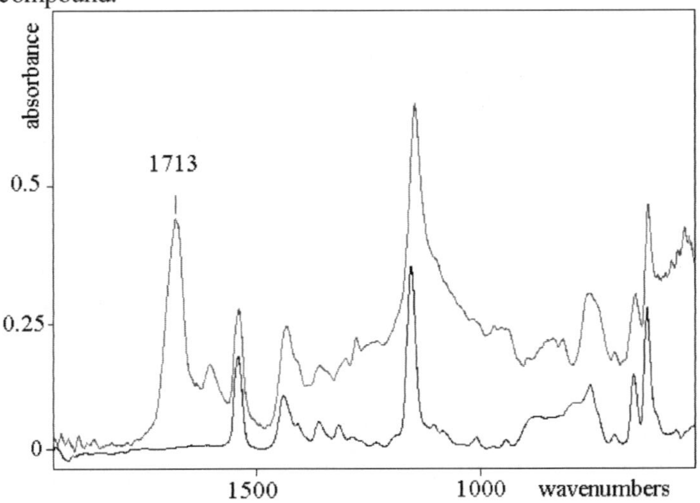

Figure 3. FT-IR of synthesized PIL.

The TLC tests were also conducted for the yielded PIL as well. Surprisingly it was observed that the IL (1-(6-hydroxyhexyl)-3-methylimiadazolium bromide) and PIL (1-[(6-methacryloyloxy)hexyl]3-methylimidazolium bromide) showed the same polarity. It was also observed that PIL and IL cannot be separated in TLC. In the Figure 3, the FT-IR spectra is represented to verify the polymerizable group formed in the synthesized PIL. The peak at 1713 cm^{-1} for PIL was assigned to the C=O stretching mode of the methacrylate group.

5. Conclusions

Synthesis the polymerizable ionic liquid 1-[(6-methacryloyloxy)hexyl]3-methylimidazolium bromide were successfully conducted. The reaction compound was yielded as transparent viscous liquid. To identify the nature of reaction compound and prove its structural evidence, several chromatographic tests were conducted followed by the principle of thin layer chromatography (TLC) and the results were discussed with relevant features and probable reasons. From the TLC results, it could be anticipated that the IL compound is polar, whereas, the starting materials are nonpolar.
The synthesis of IL was also conducted in microwave irradiation in order to reduce the reaction time. Unfortunately the reaction went not fully and side reactions were observed, thus microwave treatment necessitate individual investigation. The synthesis process was also conducted with heat treatment for 24 hours at 80 °C. The yield compound was observed as solid salt structure when the liquid comes in contact with air. The probable reason is as the oxidation of the ILs under heat.

References:

[1] Nishimura, N. and Ohno, H., (2014), “15th anniversary of polymerised ionic liquids”, Polymer, Vol. 55, pp. 3289-3297.
[2] Esko, K. P., Taaber,T., Saal, K., Lohmus, R., Kink, I. and Mäeorg, U., (2013), “New method for synthesis of methacrylate type polymerizable ionic liquids”, Synthetic Communications, Vol. 43, pp. 2846-2852.
[3] Esko, K. P., Taaber,T., Saal, K., Lohmus, R., Kink, I. and Mäeorg, U., (2013), “Increased conductivity of polymerized ionic liquids through the use of a nonpolymerizable ionic liquid additive”, Journal of Materials Research, Vol. 28(22), pp.3086-3093.
[4] Yuan, J. and Antonietti, M., (2011), “Poly(ionic liquid)s: Polymers expanding classical property profiles”, Polymer, Vol. 52, pp. 1469-1482.
[5] Fukushima, T., Kosaka, A., Yamamoto, Y., Aimiya,T., Notazawa, S., Takigawa, T., Inabe, T. and Aida, T., (2006), “Dramatic Effect of Dispersed Carbon Nanotubes on the Mechanical and Electroconductive Properties of Polymers Derived from Ionic Liquids”, Small, Vol. 2(4), pp. 554-560.
[6] Yamashita, A., Yoshio, M., Shimizu, S., Ichikawa, T., Ohno, H. and Kato, T., (2014), “Columnar Nanostructured Polymer Films Containing Ionic Liquids in Supramolecular OneDimensional Nanochannels”, Journal of Polymer Science Part A: Polymer Chemistry, Vol. 53(2), pp. 366–371.
[7] Yoshizawa, M. and Ohno, H., (2001), “Synthesis of molten salt-type polymer brush and effect of brush structure on the ionic conductivity”, Electrochimica Acta, Vol. 46(10), pp.1723-1728.
[8] Fair, J. D. and Kormos, C. M., (2008), “Flash column chromatograms estimated from thinlayer chromatography data”, Journal of Chromatography A, Vol. 1211(1–2), pp. 49-54.
[9] Baczek, T. and Sparzak, B., (2006), “Ionic Liquids as Novel Solvent Additives to Separate Peptides”, Verlag der Zeitschrift für Naturforschung, Vol. 61c, pp. 827-832.
[10] Marszałł, M. P., Sroka, W. D., Balinowska, A., Mieszkowski, D., Koba, M. and Kaliszan, R., (2013), “Ionic liquids as mobile phase additives for feasible assay of naphazoline in pharmaceutical formulation by HPTLC UV- densitometric method”, Journal of Chromatographic Science, Vol. 51(6), pp. 560-565.
[11] Habicht, J. and Mäeorg, U., (2014), “Coupling of resorcinols in retorted kukersite semicoke”, Proceedings of the Estonian Academy of Sciences, Vol. 63(1), pp. 75-85.
[12] Flieger, J., (2011), “Application of Ionic Liquids in Liquid Chromatography”, Ionic Liquids: Applications and Perspectives, book edited by Alexander Kokorin, ISBN 978-953-307-248-7, Chapter 12, pp. 243-272.

Acknowledgements.

The Archimedes Foundation Implementing Agency of Structural Support is acknowledged for support under grant no. “Carbon Nanotube Reinforced Electrospun Nano-fibres and Yarns”, no. 3.2.1101.12-0018.

IOP Conf. Series: Materials Science and Engineering 111 (2016) 012022 doi:10.1088/1757-899X/111/1/012022

Multiliposomal nanocontainers based on anionic solid liposomes and spherical polycationic brushes

AA Efimova, OV Zaborova and AV Sybachin

Department of Chemistry, M.V.Lomonosov Moscow State University, 119991 Moscow, Russian Federation

e-mail: ephimova@genebee.msu.su

Abstract. The incorporation of cholesterol in membrane of liposomes in gel state opens the gates of their usage as cargo-carrying component in complex multiliposomal nanocontainers.

1. Introduction

For last decades bilayer lipid vesicles (liposomes) have been used as nanocontainers for encapsulation and delivery of biologically active substances [1-4]. Recently in order to increase the efficacy of liposome uptake by cells and therapeutic effect of a liposomal drug it was proposed to adsorb electrostatically anionic liquid liposomes on the surface of polystyrene microspheres with grafted polycationic chains (spherical polycationic brushes) [5,6]. It is well known that the presence of unsaturated bonds in the hydrocarbon chains leads to the decrease in the thermodynamic stability of liposomes and acceleration of the processes of lipid oxidation [7-10]. In order to increase stability of liposomes in complex with polycation the fraction of unsaturated bonds in fatty tails of lipids should be drastically decreased. On the other hand increase of saturated tails in liposomes may shift phase state of lipid bilayer from liquid crystalline ("liquid") to gel ("solid"). Phase state of liposomal membrane plays significant role in formation of complexes of lipid vesicles with polymers and physico-chemical properties of the resulting compounds. In the present study we describe complexation and properties of multiliposomal nanocontainers based on solid liposomes composed of zwitter-ionic dipalmipoylphosphatidylcholine (DPPC), anionic phosphatidylserine (PS) and spherical polycationic brushes (SPB).

2. Experimental Section

DPPC, PS, N-fluoresceinisothiocyanyldipalmitoylphosphatidylethanolamine (FITC-DPPE) and cholesterol from Avanti were used as received. Small unilamellar liposomes 40-60 nm in diameter, including those loaded by NaCl solution, were prepared by the standard sonication procedure [11]. The molar ratio of PS was equal to 0.1. Liposomes with a fluorescent dye incorporated into the bilayer, were prepared by the same procedure, except 0.1 wt.% of FITC-DPPE was added to the lipid mixture before organic solution evaporation. Liposomes with sodium chloride solution in the inner water cavity were prepared via dispersion of the lipid film in a 10^{-3} M Tris buffer additionally containing 1 M NaCl. The resulting suspension was dialyzed for 1.5 h against a 10^{-3} M Tris buffer that was renewed every 30 minutes.

SPBs samples were provided by M. Ballauff, the details of synthesis and characterization of SPBs were described elsewhere [12].

Mean hydrodynamic diameters of SPBs, liposomes and SPB/liposome complexes and their electrophoretic mobility (EPM) were determined at Brookhaven Zeta Plus. The fluorescence intensity was measured using a F-4000 Hitachi fluorimeter. Permeability of the liposomal membranes towards inorganic ions was investigated by measuring the conductivity of NaCl-loaded vesicle suspensions with a CDM83 conductometer (Radiometer) as described in [13].

For the lipids and SPB structures see Supplement (S1).

3. Results and Discussion

Formation of complexes was studied as follows. A suspension of SPBs was mixed with a suspension of fluorescent-labeled FITC-DPPE liposomes, and 5 min after a SPB/liposome complex was separated by centrifugation. The fluorescence intensity of the supernatant upon concentration of liposomes added to SPBs is presented in **Figure 1**. It follows from the figure that all added liposomes were bounded to SPBs up to a concentration 1.6 mg/ml; at higher concentrations free (unbound) liposomes could be detected in the supernatant.

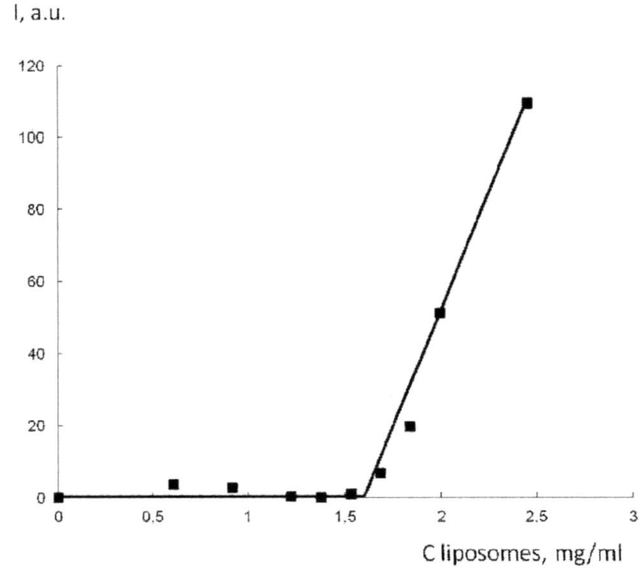

Figure 1. Fluorescence intensity of DPPC/PS liposomes in supernatant after separation of SPB/liposome complex vs total liposome concentration. ν_{PS} = 0.1; [SPB+]= 10^{-4} M; 10^{-3} M Tris buffer, pH 7.

It was reported earlier that complexation of anionic "solid" liposomes with linear polycations is accompanied by formation of defects in the liposomal membrane which initiate a leakage of the content from liposomes to surrounding solution [14] and thus restricts the use of these constructions as containers. So we controlled the integrity of liposomes complexed with SPBs by conductometry. The relative conductivity of the NaCl-loaded liposome suspension after SPB addition is shown in **Figure 2**. Complexation is followed by increase of conductivity of the system indicating that electrostatic sorption of DPPC/PS liposomes on SPBs results in formation of defects in membrane. At the same time no change in conductivity of initial DPPC/PS liposomes suspension unbound to SPBs was detected. So the modification of membrane is needed to prevent leakage of internal cargo.

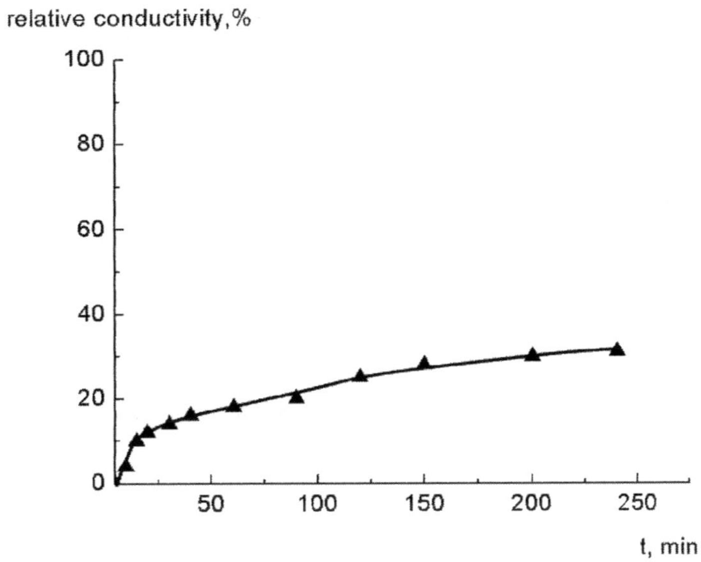

Figure 2. Time-dependent change in the conductivity of DPPC/PS liposomes, loaded with 1 M NaCl solution. ν_{PS} = 0.1; total liposome concentration 1 mg/ml; [SPB+]= 10^{-4} M; 10^{-3} M Tris buffer, pH 7. The maximum possible increase in conductivity due to the irreversible destruction of liposomes was achieved by addition of 10-fold excess of Triton X-100 surfactant.

Incorporation of cholesterol (Chol) into liposomal membrane allows one to decrease defectiveness of vesicles and prevent an early release of encapsulated drugs [15,16]. So the membrane of DPPC/PS liposomes was modified with Chol with molar fraction equal 0.2 to prevent formation of defects after complexation with SPBs.

An effect of Chol on the permeability of the liposomal membrane in complex with SPB was examined by using conductometry as described above. Time-dependent relative conductivity of the DPPC/PS/Chol liposome suspension after SPB addition was measured. No salt leakage was detected after 200 minutes of investigation. Chol incorporation into membrane led to a gradual healing of defects in the membrane. Thus, by incorporation of Chol into membrane the leakage of salt from DPPC/PS solid liposomes complexed with SPB can be completely suppressed.

The key question is whether addition of Chol to DPPC/PS liposomal membrane changes character of interaction of vesicles with SPBs or not. As electrostatic binding determines properties of the complexes between anionic liposomes and polycationic brushes the complexation of DPPC/PS and DPPC/PS/Chol liposomes with SPB was studied by measuring of electrophoretical mobility (EPM). For the both types of liposomes the addition of liposomes to SPBs resulted in decrease of surface charge. In **Figure 3** the dependence of EPM of the complexes upon concentration of vesicles is presented. The point of electroneutrality (EPM = 0) for the DPPC/PS and DPPC/PS/Chol was found to be 1.5 mg/ml. This important result indicates that incorporation of less than 0.2 molar fraction of Chol in membrane of DPPC/PS does not affect on complexation with SPBs.

We see therefore that, by Chol incorporation into the solid liposomal membrane we can control the integrity of liposomes complexed with SPBs. However the presence of Chol did not affect the possibility of the liposomal membrane to interact with oppositely charged SPBs. This result is of interest for preparing liposomal containers for drug encapsulation.

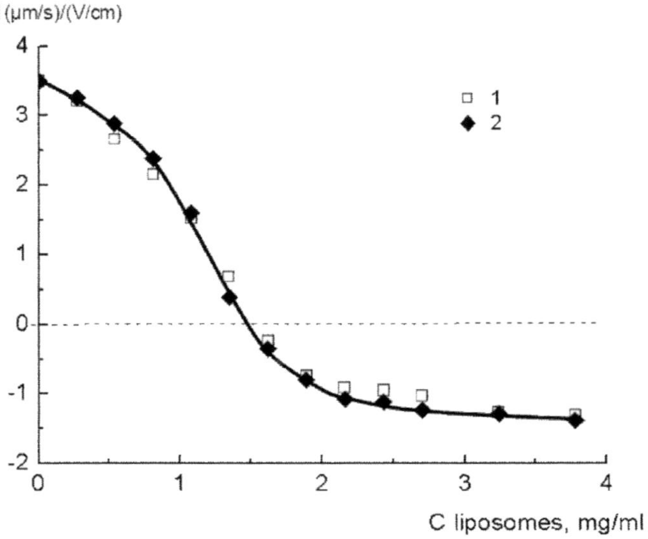

Figure 3. EPM of SPB particles vs liposome concentration for DPPC/PS (1) and DPPC/PS/Chol (2) liposomes. $\nu_{PS} = 0.1$; $\nu_{Chol} = 0.2$; [SPB+]= 10^{-4} M; 10^{-2} M Tris buffer, pH 7.

Acknowledgments
This work was supported by Russian Foundation for Basic Research (project № 15-33-20880)

References
[1] Torchilin V and Weissig W 2003 *Liposomes: A practical approach* (Oxford, Oxford University Press)
[2] Burgess P, Hutt P B, Farokhzad O C, Langer R, Minick S and Zale S 2010 *Nat. Biotechnol.* **28** 1267
[3] Immordino M L 2006 *Int. J. Nanomedicine* **1** 297
[4] Knapp C M and Whitehead K A 2014 *Exert. Opin. Drug Deliv.* **11** 1923
[5] Yaroslavov A A, Sybachin A V, Schrinner M, Ballauff M, Tsarkova L, Kesselman E, Schmidt J, Talmon Y and Menger F M 2010 *J. Am. Chem. Soc.* **132** 5948
[6] Sybachin A V, Zaborova O V, Orlov V N, Semenyuk P I, Ballauff M, Kesselman E, Schmidt J, Talmon Y, Menger F M and Yaroslavov A A 2014 *Langmuir* **30** 2441
[7] Colin A, Reggers J, Castronovo V and Ansseau M. 2003 *Encephale* **29** 49
[8] Waraho T, McClements D J and Decker E A 2011 *Food Chem.* **129** 854
[9] Ansari F A, Ali S N and Mahmood R 2015 Toxicol. In Vitro **29** 1878
[10] Johnson D R and Decker E A 2015 *Annu. Rev. Food. Sci. Technol.* **6** 171
[11] Bordi F, Cametti C, Sennato S and Viscomi D J 2006 *Colloid Interface Sci.* **304** 512
[12] Mei Y, Wittemann A, Sharma G, Ballauff M, Koch Th, Gliemann H, Horbach J and Schimmel Th 2003 *Macromolecules* **36** 3452
[13] Yaroslavov A, Efimova A and Kostenko S 2012 *Polymer Science - Series A* **54** 264
[14] Yaroslavov A A, Efimova A A, Lobyshev V I and Kabanov V A 2002 *Biochim. Biophys. Acta* **1560** 14
[15] Bouaoud C, Lebouille J G, Mendes E, De Braal H E and Meesters G M 2015 *J. Liposome Res.* **26** 1
[16] Briuglia M L, Rotella C, McFarlane A and Lamprou D A 2015 *Drug Deliv. Transl. Res.* **5** 231

Stabilizing derma collagen structure with modified dispersions of montmorillonite

M.O. Maruhlenko [1], V.A. Palamar [2], O.R. Mokrousova [2]

[1] Department of Technologies of Leather and Fur, Kiev National University of Technologies and Design, Nemirovicha-Danchenko str., 2, 01011, Kiev, Ukraine
[2] Department of Commodity Research and Examination of Nonfoods, Kiev National University of Trade and Economics, Kioto str., 119, 02156, Kiev, Ukraine
E-mail: mokrousovaolena@mail.ru

Abstract. The article reviews structural changes of derma collagen resulting from processing with modified montmorillonite dispersions. The infrared spectroscopic studies of gelatin as a simulated collagen established chemical interaction of chromium modified montmorillonite dispersions with reactive groups of the native protein. The derma collagen structure is stabilized through formation of hydrogen bonds with the surface hydroxyl groups of montmorillonite of Si-O... H-C and Si-O-H...O-C type. The formation of hydrogen bond between the nitrogen atoms of amino groups of collagen and the water proton in the interlayer space of MMT with cations in the exchange complex is possible. The formation of hydrogen bonds of Si-O....H-C type between the functional groups of chromium modified montmorillonite with CH-groups of protein, as well as Si-O....H-N with NH - groups of protein in the low-frequency spectrum range is also very likely. The efficiency of stabilizing collagen with chromium modified montmorillonite dispersions is proven by an increase of the gelatin melting temperature with mineral consumption of 3-4% of protein weight and chromium compounds (III) for the modification of the mineral around 10% of montmorillonite weight.

1. Introduction

The derma collagen is an amphoteric protein. The amphoteric nature of the collagen is caused by carboxyl, amine, peptide, hydroxyl and others groups. Thus, bonds of different nature and location can be formed in the collagen structure [1].

During the processing of animal skins into natural leather, the destruction happens between some of the bonds in the structural elements of the collagen and the formation of new bonds by reaction with chemicals [2].

For effective stabilization of the collagen structure of derma, it is necessary to achieve the structuring of its components at the levels of size. For this, it is advisable to use materials with different functional groups and different sizes of particles.

Mineral and organic tanning agents are used to stabilize the derma collagen structure. The basic chromium sulfate is used as a mineral tanning agent to produce around 80% of the leather in the world. The chromium tanning agents are complex compounds that efficiently bind the structure of collagen in various structural levels as a result of complex creation. They create a chemical bond mainly with the carboxyl groups of the collagen. The stabilization of the derma collagen structure results in the raising of collagen shrinking temperature to 100°C and the hardening of the derma structure against external impact. The main disadvantage of this type of tanning is inefficient consumption of chromium compounds, as 40% of them remains in the waste tanning solution [3, 4].

Organic tanning agents, polymeric compounds and mineral fillers (kaolin, montmorillonite, chalk, etc.) are used to bind the derma structure through other functional groups of derma collagen.

The dispersions of montmorillonite (MMT) are of practical interest because after the modification of the particle of dispersion they may change the surface charge and size. Targeted modification of montmorillonite will allow to effectively stabilizing the derma collagen.

The purpose of this work includes a study of the efficiency of structuring the derma collagen with a modified montmorillonite dispersion. In fact, the montmorillonite dispersions are planned to be modified with basic chromium sulphate (Cr-MMT). It is assumed that the use of modified montmorillonite dispersion at the stage of tanning will promote rational use of chromium compounds through the formation of bonds with functional groups of collagen and active centers of montmorillonite [5].

2. Experimental

2.1 *Materials and preparation*

Gelatin of SER "Lisichanskiy Gelatin Factory" (Ukraine) was used for the investigation as a model of derma collagen [3]. The properties of gelatin (GOST 11293-89): particle size <5 mm; pH in a soluble state in the amount of 1.0% = 5,6 ± 0.1%, protein content = 83.32 ± 0.03%; ash = 1.38 ± 0.01%; moisture content = 15.3 ± 0.1%.

The basic chromium sulfate ($Cr_2(SO_4)_n(OH)_{6-2n}$) with basicity 40.0% (Kazakhstan, Aktiubinsk plant of chromium compounds) was used as a chromium tanning agent. The content of chromium oxide (III) Cr_2O_3 - 25.6%.

Bentonite clays of Dashukovsky deposit (Ukraine) was used for investigation. The content of montmorillonite in bentonite rocks constituted 85%.

Montmorillonite dispersion was modified in stages by sodium carbonate and basic chromium sulfate. For that, sodium carbonate in a soluble state in the amount of 6.0%, reported at mineral weight, was introduced into water dispersion with the montmorillonite concentration of 100 g/l. The obtained dispersion was mechanical mixed for 120 minutes. Then, basic chromium sulfate in a soluble state in the amount of 10.0% Cr_2O_3, reported at mineral weight was introduced to the Na-montmorillonite dispersion.

As the objects for FTIR spectroscopy investigation, we used the film of gelatin (G), of chromium - tanned gelatin (G_{Cr}) and Cr-MMT tanned gelatin ($G_{Cr\text{-}MMT}$).

The chromium - tanned gelatin (G_{Cr}) was prepared as follows: the solution (5%) of eatable gelatin was treated during the 3 hours with chromium (III) salts in the quantity of 3.0% Cr_2O_3 counting on dry protein at the temperature of 20 ± 2°C. pH was 3.5.

The Cr-MMT added into the gelatin solution in the quantity of 6% reported at dry substance at the temperature of 20 ± 2°C. pH was 3.7. The films with the thickness of 100 μm were formed by pouring on the Teflon disks, with following drying at room temperature for 72 hours, and then in the oven at 30° C till constant weight.

2.2 *Characterization techniques*

The IR spectroscopic studies were performed on the universal Fourier transform infrared spectrometer TENSOR-37 (BRUKER, Germany) within the 4000-400 cm^{-1}. The nature and efficiency of interactions of functional groups of montmorillonite with active groups of gelatin were evaluated by the change in intensity of optical densities of the corresponding characteristic absorption bands. The absorption bands in the infrared spectra of precursors and products of their interaction were interpreted using data of literature [6-10] in accordance with the frequencies of the characteristic groups of atoms. The spectra of the tested samples were used to calculate the peaks areas of relative absorption densities by the Gauss equation [11]. The changes of configuration of absorption bands and their shifts allow drawing a conclusion about the nature of intermolecular interactions in tanning collagen with chromium tanning agents.

The melting point was determined for the galantine of native gelatin (G), the gelatin treated with chromium-modified montmorillonite dispersions ($G_{Cr\text{-}MMT}$) at a constant consumption rate of the mineral or chromium oxide. The melting temperature for gelatin was determined by the fusiometric method (GOST 11293-89). The method is driven by determining the flowable temperature of gelatin galantine. The gelatin solution with the concentration of 2.5% was prepared for the research and transferred to a test tube. Then, chromium-modified montmorillonite dispersion was added to gelatin and the tube was left alone for 24 hours. Upon complete solidification of gelatin galantine, a metal ball was placed on the surface and the test tube was immersed in a glycerin solution. The glycerin solution was gradually heated at the speed of 3-5 °C / min. The temperature at which the ball fell on the bottom of the tube indicated the galantine melting point.

3. Results and discussion

To evaluate stabilization of the derma collagen structure as a result of chemical interactions between the active groups of collagen and tannins, the information about the characteristic absorption bands of the input substances presented in Fig. 1 and Table 1 was used.
The most characteristic gelatin absorption bands are observed in the frequency range of 3300-3100 cm^{-1} and 1650-1245 cm^{-1} (Fig. 1, Table 1).

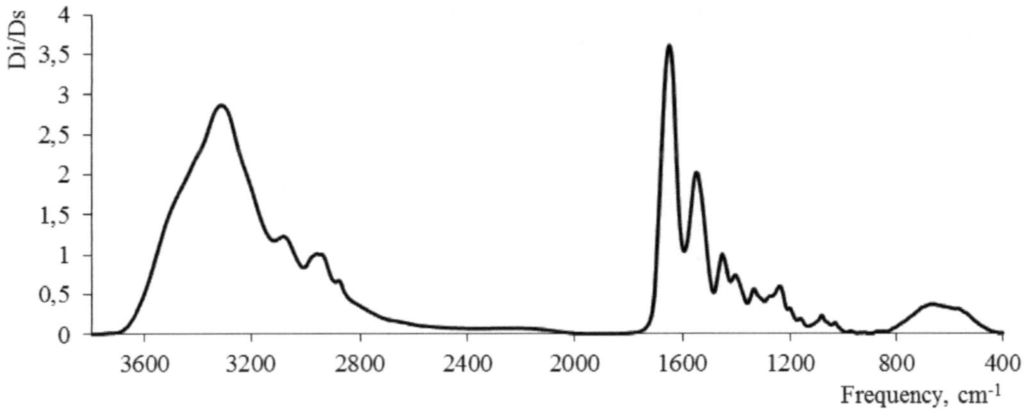

Fig.1. Characteristic absorption bands of gelatin

Table 1. Characteristic bands of gelatin

Gelatin			
Band (cm^{-1})	Group	Band (cm^{-1})	Group
3330, 3200 -3500	- NH_{bound}, - OH	1335	Amid III, -CN + -NH
3071	- NH_{bound}	1245	-CN, -NH
2955	- CH_2 asym	1162	$NH^{3}{}_{+}$
2834	- CH_2 sym	1082	- CN
1654	Amid I, -C = O	1031	- C-O-C = C
1543	Amid II, -NH, -CN	972	-NH_{bound}
1454	CH_2	672	- NH_{bound}, Amide V,
1398	- C = O (COO^-)	549	Amid V I, -NH, -C = O

The first area corresponds to valence fluctuations of the associated NH_2, NH and OH-groups. The band at 3330 cm^{-1} is due to the bonded groups of NH (Amide A), that participate in the creation of intramolecular hydrogen bonds. The bands at 2954 cm^{-1}, 2834 cm^{-1} are associated with the asymmetric and symmetric stretching vibration of CH_2 groups. In the area of 1650 - 1234 cm^{-1}, absorption bands of Amide I, Amide II and Amide III are observed. This area is associated with the absorption band of carbonyl groups C=O and deformation vibrations of NH-groups, including amine, and imine guanosine groups. The intense band of 1650-1660 cm^{-1} is due to valence fluctuations of associated carbonyl groups C=O (Amide I). The band 1540-1550 cm^{-1} characterizes the deformation vibrations of NH-groups (Amide II) and, to a lesser extent, those of CN-groups. The Amide II band is slightly weaker from the carbonyl absorption bands. The interval of 1197-941 cm^{-1} combines a set of bands. The 1163 cm^{-1} band indicates the pendulum oscillations of NH_3^+ groups, 1082 cm^{-1} and 1031 cm^{-1} bands are characteristic for valence groups CN, C-O and C=C, and the band at 972 cm^{-1} corresponds to the bound NH - groups. In the 840-400 cm^{-1} area, characteristic deformation vibrations of bound NH-groups with peaks of 672 cm^{-1} (Amide V) and 549 cm^{-1} (Amide VI). These bands are wide and less intense.

The spectrum of montmorillonite (Fig. 2) is characterized by more intense absorption bands in both high-frequency and low-frequency zones. The strip stretching OH vibrations that are characteristic of structural hydroxyl groups of layered silicates are observed at 3700-3200 cm^{-1}. The weak band with frequency vibrations of 1637 cm^{-1} probably characterizes the deformation vibrations of the H-O-H groups. Also, the intense band at 1168-1046 cm^{-1} which is characteristic of silicon oxygen structures, less intense band at 788 cm^{-1} and strong absorption bands at 529 -461 cm^{1} (Fig. 2) are observed. The pronounced wide band (Fig. 2) at 1042 cm^{-1} corresponds to valence fluctuations of asymmetric and symmetric Si-O-Si tetrahedral of silicon-oxygen frame and indicates presence of surface hydroxyl groups. The band in the range of 788 cm^{-1} corresponds to fluctuations of AlO_4 in the tetrahedral rings of (O-Si-O). The band in the area 694 cm^{-1} is characteristic of structural deformation vibrations of hydroxyl group of the mineral that are associated with octahedral Al^{3+} cations. The intensive doublet at 463 and 529 cm^{-1} is characteristic of dioctahedral minerals of the montmorillonite type and indicates the deformation vibrations of links Me-O and (O-Si-O).

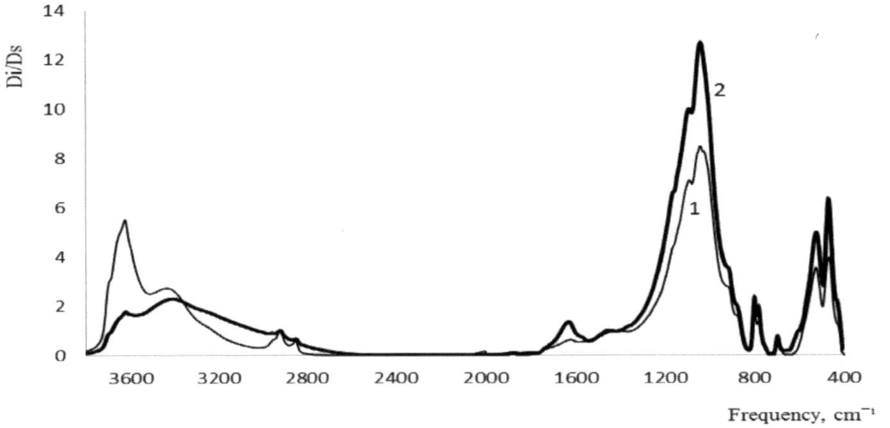

Figure 2. Characteristic absorption bands of montmorillonite (1) and Cr-montmorillonite (2)

The modification of montmorillonite with basic chromium sulphate promotes change of the characteristic absorption bands of high and low-frequency range zones of the mineral (Fig. 2, Table 2). For Cr-MMT, the band of bound absorbed water at 3454 cm^{-1} is shifted to lower frequencies (3428 cm^{-1}). There is a decrease in its intensity due to OH stretching bands.

The major changes in the area of high-frequency spectrum Cr-MMT are associated with the redistribution of water in the MMT interburst space. The deformation fluctuations in H-O-H groups for the range of Cr-MMT are observed at 1642 cm^{-1} and are characterized by 7 times the intensity compared with the MMT. For Cr-MMT band of 3192 cm^{-1} shifted to 260 cm^{-1} in a high-frequency region compared to the band at 2932 cm^{-1} for the MMT spectrum. Apparently, the phenomenon is due to hydrogen bonds. The chromium complex displaces water from the interlayer space of the MMT to form hydroxocomplex ions, which have a high polarization and adsorption capacity. The presence of the Cr^{+3} hydroxocomplex ions between the silicate layers of montmorillonite is manifested by the siloxane groups growing around 1046 cm^{-1} and shrinking OH groups. That also results in almost doubling the areas under the respective peaks. Such changes may indicate a strengthening MMT structure.

The peaks at 529 and 461 cm^{-1} that correspond to valence fluctuations of the Cr-O bond in the infrared area have a slightly larger area under the peaks compared with the MMT spectrum, indicating a more orderly crystal structure of Cr_2O_3 on the MMT surface.

Table 2. Characteristic bands of modified montmorillonite and change of their optical densities

Band (cm $^{-1}$)	Group	Optical density, D/D_0 (for peak areas)	
		MMT	Cr –MMT
3640-3641	- OH $_{free}$	457.8	113.0
3454-3428	- OH $_{bound}$	980.0	754.9
2932-3192	- OH	91.3	169.1
2854-2980	H_2O, - OH	6.7	295.1
1637-1642	- (H-O-H)	12.8	91.4
1168-1270	Si-O	207.0	413.3
1131-1189	Si-O	537.5	307.6
1042-1046	n(Si-O-Si), AlO_4 – - tetrahedral	1225.5	2028.7
788	- (O-Si-O), AlO_4	36.0	76.9
694	- (Al-OH)	14.6	12.0
529	- (O-Si-O) Me-O	205.2	357.8
463-461	- (O-Si-O) Me-O	158.2	218.4

The use of tannins based on basic chromium sulphate, including dispersions of Cr-MMT to structure the derma collagen changes the range of gelatin (Fig. 3, Table 3).

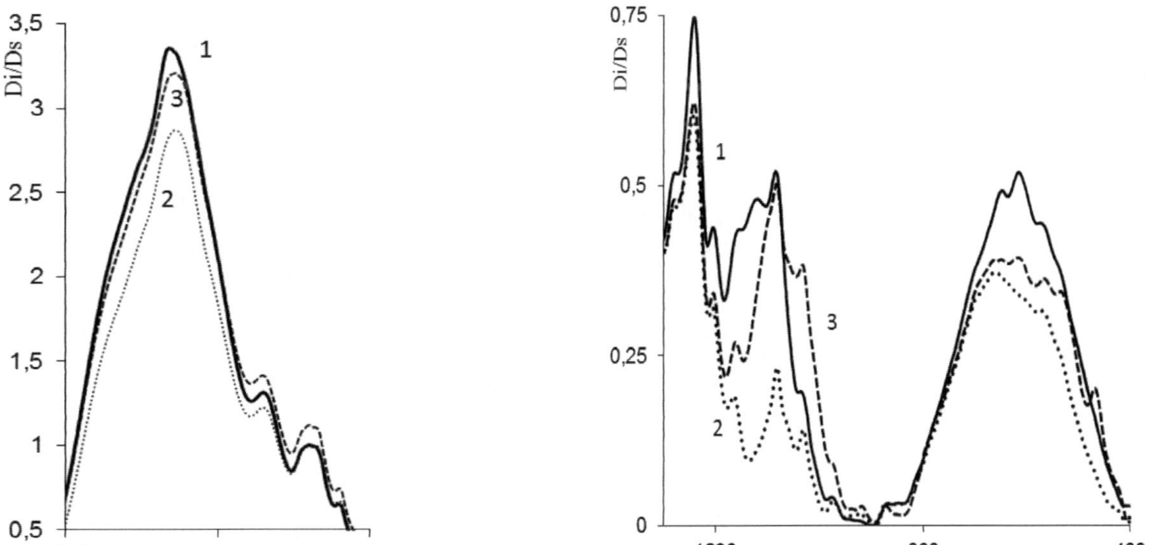

Figure 3. Characteristic absorption bands G(1), G_{cr} (2) and $G_{cr\text{-}MMT}$ (3)

Treating of gelatin with basic chromium sulphate (G_{Cr}) promotes changes of absorption bands (Fig. 3, Table 3), which are characteristic of absorption of carboxyl, amine, and OH groups. A decrease of the area under the peaks at 3333, 3072, 1082cm^{-1} and the corresponding increase of the area under the peaks of 1654, 1545, 1398, 1336, 1246 cm^{-1} can be observed in the G_{Cr} spectrum as compared to gelatin spectrum. It can be seen the emergence of an intense absorption band at 1122 cm^{-1}. This is evidence of the active interaction of carbonyl and carboxyl groups of collagen with the complex chromium making metal ion. Cross-links that lead to stabilizing of derma collagen structure are formed in the process.

The $G_{Cr\text{-}MMT}$ spectrum is characterized by absorption bands similar to the G_{Cr} spectrum, which indicates the identical mechanism of derma collagen structurization. The main changes are associated with the intensity of the absorption bands and the corresponding peak areas. The $G_{Cr\text{-}MTT}$ spectrum is characterized by a slight decrease of the area under the peaks of the absorption bands at 3341 cm^{-1} and a shift of the bands to the high-frequency region compared with the G_{Cr} spectrum.

In the range 1246 cm^{-1} ... 918 cm^{-1} for the $G_{Cr\text{-}MMT}$ significant differences are observed in the areas of the peaks corresponding to fluctuations of carbonyl, ammonium, and simple ester sulphate groups of gelatin, as well as free surface hydroxyl groups Si-O - fluctuations. The strip stretching vibrations of Amide I, Amide II and Amide III for $G_{Cr\text{-}MMT}$ show a special impact of Cr-MMT on the derma collagen structure versus G_{Cr}. The area under the peaks at 1654 cm^{-1}, 1545 cm^{-1} and 1336 cm^{-1} for $G_{Cr\text{-}MTT}$ is slightly lower compared with the peaks in the G_{Cr} spectrum. This corroborates the formation of a hydrogen bond between the nitrogen atoms of amino groups of gelatin and a water proton in the interlayer space of MMT with monovalent cations in the exchange complex. A part of gelatin molecules enters in a specific interaction with exchange cations of the mineral and simultaneously forms hydrogen bonds with the surface atoms of oxygen or adjacent atoms of silicate.

Table 3. Change of optical densities in structured gelatin spectra

Band (cm^{-1})	Group	Optical density, D/D_0 (for peak areas)		
		G	G_{Cr}	G_{CrMMT}
3330-3337-3341-3342	- NH, -OH	986.7	1198.9	1139.5
3072	NH_2, -NH	34.7	30.9	27.2
2834-2813-2824-2824	- CH_2 sym., -OH	78.1	89.3	81.1
1654	- C = O -(H-O-H)	180.1	184.4	174
1545	- NH_2	67.5	70.8	66.6
1398	- C = O (COO^-)	6.8	7.5	5.4
1336	- CN + -NH	5.7	6.4	5.2
1246	- CN, -NH	15.5	17.1	24.9
1162-1160-1145-1150	NH^{3+}	1.0	2.2	17.9
1122	S = O, -CN, C-O, SO_4^{2-}	-	12.7	-
1082-1081-1100-1103	C-O, -CN, SO_4^{2-}	16.7	10.8	4.7
1021-1034-1036	C-O, $_n$(Si-O Si), AlO_4 - tetrahedrons	2.9	3.1	88.5
972-973	- CH, NH $_{bound}$	0.4	0.3	-
918	Si-O, OH	-	-	2.1
873	CO_3^{2-}, AlO_4 -tetrahedral	-	-	1.2
786	- (O-Si-O), AlO_4	-	-	5.9
672 -685-680-691	NH, Amide V, (Si-O $_4$, Cr^{+3}, Al^{3+}	57.9	43.4	39.1
604	Amid V, -NH, Cr^{+3}	-	22.8	-
549-537-529-561	Cr-O, Amide V, -NH	23.1	42.9	105.9
524	- (O-Si-O) Me-O	-	-	14.3
466	- (O-Si-O) Me-O	-	-	14.2

There is also a decrease of the area of maximum absorption peaks at 1398 cm^{-1} for $Cr_{G\text{-}MMT}$. The spectral changes of this type are caused by intermolecular interactions where the carboxyl group acts as an electron donor. The lower peak area indicates that during tanning the binding energy goes up, i.e.

it forms a stronger bond with the carboxylic group than the chromium sulfate complex. Also, the interaction of the carboxyl group of gelatin with cations of lateral sides of the mineral surface is possible.
Based on the comparative analysis of the relative densities of absorption spectra for the $G_{Cr\text{-}MMT}$ at 2824 cm^{-1} and 680-466 cm^{-1}, we conclude that, most likely, the hydrogen bonds of type Si-O .... H-C type are formed between the functional groups Cr-MMT and CN protein groups, as well as Si-O....H-N with NH protein groups. The increased optical density of absorption bands 529-561 cm^{-1} and the appearance of the peak at 466 cm^{-1} might be the evidence of the formation of such bonds.
The analysis of the melting point of a native gelatin galantine before and after treatment with Cr-MMT dispersions showed their efficiency to structure the derma collagen.
It was found that the melting temperature and consumption of chromium tanning agents and montmorillonite are interdependent. The output of studies of the impact of consumption of chromium tanning agents (in terms of chromium oxide) and montmorillonite on the melting point of gelatin galantine is shown in Fig. 4.

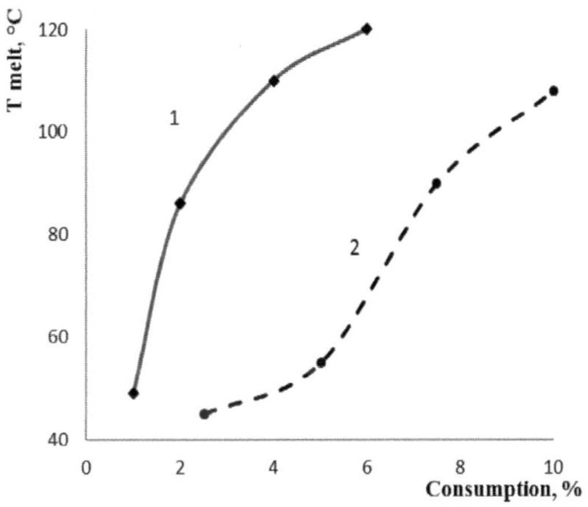

Figure 4. Dependence of chromium compound consumption (1) of MMT weight and modified montmorillonite dispersions (2) of protein weight

1 - $G_{cr\text{-}MMT}$, consumption of MMT - const (4% of protein weight)
2 - $G_{cr\text{-}MMT}$, consumption of Cr_2O_3 - const (10% of protein weight)

As shown in Fig. 4, with the increasing consumption of chromium compounds being part of Cr-MMT dispersions, one can observe a steady rise of the gelatin melting point. This results from stabilizing collagen upon its interactions with chromium complexes. In this study (Fig. 4), consumption of montmorillonite was fixed as 4% of the protein weight. In using G dispersions of Cr-MMT for treatment with consumption of 2-5% of the protein weight, the melting point goes up rather slowly. But raising the consumption of chromium oxide up to 10% as part of chromium mineral dispersion of the protein weight promotes a sharp rise in the melting point from 55 to 110°C.
The effect of montmorillonite included in Cr-MMT is most evident where the mineral consumption is at 4% of the protein weight . In this study (Fig. 4), consumption of montmorillonite was fixed as 10% of the protein weight.
Thus, according to the received output, the optimum consumption rate for efficient stabilization of derma collagen structure is about 3-4% of the protein weight, and 10% of chromium compounds from the mineral weight.

4. Conclusions

1. The article reviews efficiency of derma collagen stabilization with modified montmorillonite dispersions.

2. The infrared spectroscopic studies show that the most characteristic absorption bands of gelatin as a simulated derma collagen correspond to carboxy-, amino-, imino-, hydroxyl groups etc. Modification of montmorillonite with basic chromium sulphate promotes changes in high-frequency and low-frequency ranges of the mineral. The chromium complex displaces water from the montmorillonite interlayer space to form hydroxo-complex ions, which have high polarization and adsorption capacity to support interaction with derma collagen.
3. It is shown that the interaction of the functional groups of native gelatin with active centers of chromium modified montmorillonite is due to the formation of hydrogen bonds O-C-O and CH2 protein groups with the surface hydroxyl groups of montmorillonite of Si-O...H-C and Si-O-H... O-C type. The formation of a hydrogen bond between the nitrogen atoms of gelatin amino groups and water proton in the interlayer space of montmorillonite with cations in the exchange complex is possible. A part of gelatin molecules enters in a specific interaction with mineral exchange cations and simultaneously forms hydrogen bonds with surface oxygen atoms, or neighboring atoms of silicate that efficiently stabilizes the collagen structure. In the low-frequency spectrum range, the formation of hydrogen bonds of Si-O...H- C between the functional groups of chromium modified montmorillonite with CH- groups of protein, as well as Si-O.... H-N with NH - groups of protein is most likely. The presented structural changes are accompanied by changes of the peaks area of the characteristic absorption bands of gelatin and their shift.
4. It was found that the efficiency of stabilizing the derma collagen is manifested in the rise of the melting temperature, which is interdependent on consumption of montmorillonite and basic chromium sulphate for its modification.
5. According to the findings, the optimal consumption rate of the mineral for efficient stabilization of derma collagen is around 3-4% of the protein weight, and that of chromium (III) - 10% of the mineral weight.

References

[1] Cot J 2004 An imaginary journey to the collagen molecule for a better understanding of leather waste treatments *J. Amer. Leather Chem. Ass.* **99(8)** 322-351.
[2] Maffia G 2004 Collagen Processing *J. Amer. Leather Chem. Ass.* **99(4)** 164-170.
[3] Zhongkai Z 2006 Physicochemical Properties of Collagen, Gelatin and Collagen Hydrolysate from Bovine Limed Split Wasters *JSLTC* **1(90)** 23-29.
[4] Gaidau C 2013 Applicative Chemistry of Tanning Metallic Heterocomplexes *Benthan Science Publishers* 130 p.
[5] Mokrousova O 2015 Resources-saving Chromium Tanning of Leather with the Use of Modified Montmorillonite *Revista de chimie* **66** 353-357.
[6] Nakanishi K 1965 Infrared spectra and structure of organic compounds *M. Mir.* p. 216.
[7] Bellamy L 1974 New data on the infrared spectra of complex molecules *M . Mir.* p. 444.
[8] Lazarev A N 1968 Vibrational spectra and structure of silicates *L. Nauka*, p. 345.
[9] Chirgidze Y N 1965 Infrared spectra and structure of polypeptides and proteins *M. Nauka* p. 136.
[10] Brown D, Sainsbury M 1992 Spectroscopy of organic materials *M. Mir* 300.
[11] Kazitsyna LA 1971 Application of UV, IR, NMR spectroscopy in organic chemistry *M . High School* p. 264.

IOP Conf. Series: Materials Science and Engineering **111** (2016) 012024 doi:10.1088/1757-899X/111/1/012024

APPLICATION OF ADVANCED POLYMERIC COMPOUNDS FOR DEVELOPMENT OF LEATHER PRODUCTION

A Nikonova[1], O Andreyeva[1], L Maistrenko[1]

[1]Kyiv National University of Technologies and Design, str. Nemirovich-Danchenko, 2, Kyiv, 01011, Ukraine

E-mail: nikonovaav@mail.ua

Abstract. Abatement of negative impact of leather industry on the environment is associated with scientifically grounded use of chemical materials and raw materials, with the development of low-toxic and low-waste technologies. A solution of this problem can be development and implementation of pre-tanning and tanning processes alternative technologies with the use of polymeric compounds. To replace traditional pickling process advanced polymeric material based on acrylic acid has been used. It has been found experimentally that application of the polymeric processing intensifies the tanning process, increases the thermostability of leather with more rational use of chemical materials. The content of chrome (III) compounds, chlorides and sulphates in waste water are reduced.

1. Introduction

One of the most important processes of natural leather production is tanning, during which the main formation of the structure and many properties of derma takes place. More than 80% of the leathers in the world are processed by using compounds of chrome which is connected with easiness of technology, reliability of the process, high technological and operational properties of chrome tanned leathers, the possibility to manufacture products of different kinds. Using of known production technologies have negative impact on the environment because of formation of waste waters that contain toxic compounds of chrome (III), sulphates, chlorides, products of protein degradation etc. Depending on the technology of the production nearly 4.0-8.0 g/l of chrome is contained in waste solutions after the process of chrome tanning. At the same time the effectiveness of tannins using is nearly 50-60%, and the rest is transferred to waste waters and solid production wastes [1-2]. In addition to carrying out the process of pickling before tanning is increases concentration of sulphates in the waste solutions, which together with other salts present danger are formation of hydrogen sulphide and acid. Hereupon, the purification of drains by using biological methods becomes complicated and needs additional expenses [3].

Manufacturers try to remove the above mentioned drawbacks at the expense of partial replacement compounds of chrome by other mineral and organic tanning agents, optimization of tanning parameters according to the index of chrome exhaustion, repeated using of tanning solutions, and introduction of masking and alkali reagents. Herewith, it is not always possible to conduct the process in the way when during rational using of material and energetic resources the finished leather was of high value. The development and introduction of new technologies into production which decrease negative impact, presents scientific and practical interest. One of the most perspective directions of formation of such technologies is using of available, nontoxic chemical materials, which are able to increase considerably the diffusion component of tanning process and provide maximum fixation of chrome tanning agent by the leather tissue [4].

The previous work shows that the using of polymeric material on the basis of maleic acid before chrome tanning the increases interstructural distance in the collagen. This improves the diffusion of the tanning agent into the derma and as a result of its interaction with collagen the shrinkage temperature is raise. Duration of tanning process is decreases in 1.2 – 9.0 times. The processes of polymeric processing and chrome tanning are carried out non-traditionally at temperature 36-38°C [5].

The aim of this research is to study the influence of advanced polymeric material on the bases of acrylic acid (PMAA) on the process of chrome tanning with purpose to define the effectiveness of its using in leather production. In order to decrease harmful impact on the environment the possibility to exclude the technological cycle of pickling process was investigated.

2. Materials and methods

2.1. Raw and chemical materials

For all groups bated pelt of sheep skin (90 sq.dm area) was used.

Polymeric material PMAA (Codyeco S.p.A., Italy) was applied for experimental groups. It is one of commercial technical materials for leather production on the world market. The specification of PMAA used in the present study: solubility in water – soluble; active agent content (dry residue) – 32.5%; pH of 10% solution – 5.5±1.0; resistant to electrolytes action; ability to interact with the collagen [6].

Sodium chloride (97%), sulphuric acid (96%), formic acid (85%), chrome tanning agent (38.0% basicity, 25.0% of chrome oxide), sodium hydrocarbonate (96%) for control group and the same chemical material except of sulphuric and formic acids for experimental groups were used in the present study. All indicated materials for leather processing are the technical products. Materials for chemical analysis are the analytical grade.

2.2. Experimental conditions

Bated pelt from sheep skin were divided into 14 groups and completed in two samples in group by the method of asymmetric trimming [7]. The samples were prepared in size 10x20 cm and the edge portions was removed. Samples of pelt control group 14c was fully treated using traditional chrome-pickling tanning at temperature 18-20°C, known as production technology of garment leather [8]. All processes before pickling in the experimental groups 1-13 were performed using typical method, by without process of pickling. Polymeric processing was performed directly before tanning, bath was poured. Consumption of PMAA was in amount 1.0-3.0%, temperature 36-38°C, process time 1.0-3.0 hour (h) and consumption of water in amount 100%. The pH after polymeric processing was 6.0±0.2. Then the solution was drained.The next processing was carried with sodium chloride during 15 min to prevent acid swelling. Consumptions of sodium chloride was in amount 5%, water in amount 100% Chrome tanning agent was added to the same solution in amount 1.2-1.6% (% of chrome oxide) and temperature 36-38°C (for traditional chrome-pickling tanning the processing temperature is 20°C). After 1.0h after beginning of tanning, sodium hydrocarbonate was added in amount 0.3% (1:20 w/v) in few feeds and section pH was controlled. In all cases, the end of the tanning process was determined by a positive boiling test. Consumption of chemical materials and water is depending on limed pelt mass of samples. Conditions of polymeric processing and tanning are presented in table.1.

All processing was carried out in a laboratory setting for shaking glass containers of size 1.0 liter were used, providing required temperature and shaking. The frequency of shaking was 10 min^{-1}.

Table1. Conditions of polymeric processing and tanning.

Parameters	Group													
	1	2	4	3	5	6	7	8	9	10	11	12	13	14c
PMAA (%)	2.0	2.0	2.0	2.0	3.0	1.0	3.0	1.0	3.0	1.0	3.0	1.0	2.0	-
Chrome tanning agent (%)	1.6	1.2	1.6	1.2	1.4	1.4	1.4	1.4	1.6	1.6	1.2	1.2	1.4	1.6
Time of polymeric processing (h)	2.0	2.0	1.0	1.0	2.0	2.0	1.0	1.0	1.5	1.5	1.5	1.5	1.5	-

For determining of the parameters of non-pickling chrome-polymeric tanned (NPCPT) process in all existing methods of experimental planning was chosen widespread in practice no-compositional symmetric quasi-D-optimal plan that allows getting a good statistical model which has the minimal number of experiments. The factors influencing this method vary at three levels: +1, 0, –1 [9].

After a series of previous experiments the three most important factors are determined: X_1 – consumption of polymeric material, %; X_2 – consumption of chrome tanning agent, %; X_3 – time of polymeric processing, h, along with zero level and interval of variation, that is shown in table 2.

Table 2. Zero level and variation interval of factors.

Factor	Consumption (%)		Time of polymeric processing (h)
	PMAA	Chrome tanning agent	
Coded designations	X_1	X_2	X_3
Zero level x_{i0}	2.0	1.4	1.5
Range of variation Δx_i	1.0	0.2	0.5

According to primary variables (response function) values of tanned leather that efficiency describe tanning process ware selected: the mass part of volume yield, shrinkage temperature, mass part of chrome oxide (equation 2-4).

The necessary calculations and visuals construct the optimal area of NPCPT process was fulfilled using MathCAD 15 [10].

2.3 Research methods

Tanned leather and exhaustion solutions were subjected to physical and chemical analysis.

The pH control of solutions was determined using a pH-meter PATECH PH-013M.

Shrinkage temperature, which is the measure of degree of tanning, was analysed for experimental and for control tannned leathers. The measurements were carried out using a shrinkage tester as per official testing method (ISO 3380:2015). Boiling test was carried out in accordance [7].

A full penetration of tanning agent into derma was measured performed at intervals of 30 min using Bresser Researcher Bino optical microscope with increasing 40-1000 (Bresser, Germany).

Phisical and chemical analysis of leather ware carried out as per official methods: mass part of chrome oxide ISO 5398-1:2007, moisture content ISO 4684:2005, hide substance ISO5397:1984), apparent density ISO 2420:2002, thickness ISO 2589:2002.

Tanned solutions was analysed using a photocolorimeter AE-30F (ERMA Inc, Japan) for determination of chrome content (Cr_2O_3) as per official analytical method ISO 5398-2:2009.

Collecting and analysis of waste waters was made as per official methods of chemical analysis for BOD ISO 5815-1:2003, ISO 5815-2:2003, COD ISO 15705:2002, sulfates content ISO 6844:1983, chlorides content ISO 3566:1976.

Volume yield VR, cm^3/100g of protein was calculated using the formula (1) [11]:

$$VR = \frac{100^2 \cdot d_a}{HS} \tag{1}$$

where d_a is the apparent density, g/cm^3; HS is the hide substance, (per absolutly dry substance), %.

Leathers after tanning were subjected to visual and feel assessment for functional properties such as softness, fullness, grain smoothness characteristics and general appearance by hand and visual examination. The absence of complications should be noted during processing of experimental groups the samples are soft and fulling, have a clean and silky grain smoothness.

3. Results and discussion

Adequate mathematical models in the encoded units were obtained. These models describe the dependence of the most important indices of leather after NPCPT process of expenses of chrome tanning agent, PMAA and time of polymeric processing.

The mathematical on down.

- *Volume yield:*

$$Y_1 = 225.0 + 9.81X_2 + 7.75X_1X_2 + 19.02X_1^2 + 9.2X_3^2 \quad (2)$$

- *Shrinkage temperature:*

$$Y_2 = 108.5 - 0.75X_1 + 1.56X_2 + 1.75X_1X_2 - 1.75X_1X_3 + 0.81X_1^2 + 0.94X_2^2 \quad (3)$$

- *The mass part of chrome oxide:*

$$Y_3 = 6.22 + 0.35X_1 + 0.19X_2 - 0.23X_3 + 0.17X_1X_2 - 0.29X_1X_3 - 0.24X_2X_3 - 0.41X_1^2 + 0.85X_2^2 - 0.36X_3^2 \quad (4)$$

For full description of tanning process in choosing the rational parameters of NPCPT process the several factors were tested additionally: penetration of tanning agent, duration of tanning, exhaustion of tanning agent, thickness and volume yield (table 3).

The experimental results shows that in comparison with the traditional chrome-pickling tanning at temperature 18-20°C, processing with PMAA at temperature 36-38°C improved next parameters: penetration of tanning agent into derma increased in 0.7-3.0 times, duration of tanning process reduced to 2.5-5.0 times, exhaustion of chrome tanning agent from solution increased to 13.5-38.6%.

The use of PMAA over 2.0h has no effect on exhaustion of chrome tanning agent from working solution that is at the level 62.9-70.0%. At lower time of polymeric processing (1.0h) the exhaustion of tanning agent better on 3.7-18.6%. Best rate (88.6%) was obtained for polymeric processing during 1.0h and consumption 1.0% of PMAA. Regardless of exhaustion of chrome tanning agent (79.2-88.6%), the best exhaustion of tanning agent was achieved at lower consumption of PMAA (1.0%) and less time of polymeric processing (1.0h), or higher consumption of PMAA (3.0%) and less time of polymeric processing (1.0h). In other conditions tanning processes rate decreased at 11.5-25.7%.

Table 3. Rational parameters choosing of NPCPT process.

Group	Full penetration of tanning agent (h)	Duration of tanning (h)	Exhaustion of tanning agent (%)	The mass part of chrome oxide (%)	Shrinkage temperature (°C)	Volume yield (cm^3/100g)	Yield of thickness relative to bated pelt (%)
1	0.5	3.0	65.0	6.6	111.0	238.9	80.0
2	0.5	3.0	70.0	6.5	106.5	207.2	50.1
4	1.0	4.0	70.0	7.4	113.0	257.9	88.7
3	1.0	4.0	66.6	6.4	107.0	233.7	76.7
5	1.0	3.0	62.9	5.2	108.5	258.6	108.8
6	0.5	3.0	64.3	5.1	111.0	278.5	77.4
7	1.0	4.0	78.6	6.4	111.0	239.2	61.2
8	1.0	4.0	88.6	5.1	106.5	236.6	66.0
9	1.0	3.0	70.0	7.3	110.5	255.8	128.3
10	0.5	2.0	71.9	6.3	111.0	244.0	115.1
11	1.0	4.0	79.2	6.7	106.0	229.0	106.3
12	1.0	4.0	69.2	6.4	113.5	248.2	98.3
13	1.0	4.0	77.1	6.2	108.5	225.0	91.1
14c	1.5	10.0	54.4	3.1	97.5	237.0	93.0

IOP Conf. Series: Materials Science and Engineering **111** (2016) 012024 doi:10.1088/1757-899X/111/1/012024

The visuals construct the optimal area and conditionally optimal values of NPCPT leather was intended using mathematical equations and presented on figure 1.

Figure 1 show that optimal parameters of NPCPT process by following ranges of coded units (x_1 = -0.8; x_2 = +0.7; x_3 = -0.4), represent the following functions report: volume yield Y_1 = 241.7 $cm^3/100g$, shrinkage temperature Y_2 = 109.6°C; the mass part of chrome oxide Y_3 = 6.1%.

Subsequently, on the set of all parameters that describe the processing efficiency 10 group was chosen, in which the processing parameters (consumption of PMAA in amount 1.0% and chrome tanning agent in amount 1.6%, polymeric processing at 1.5h) provide appropriate level of exhaustion tanning agent from working solution, high mass part of chrome oxide in tanned leather, high shrinkage temperature, thickness and volume yield.

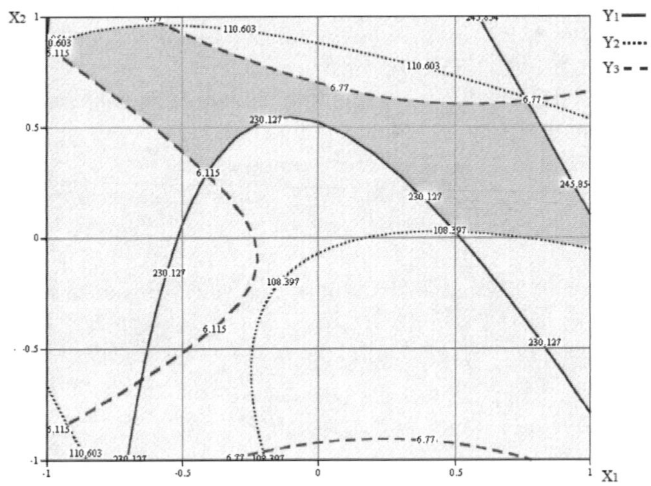

Figure 1. Diagram of optimal area of NPCPT process.

As compared with the known technology garment leather production that provides chrome-pickling tanning, new technology excludes pickling process and decrease the environmental and energy impact on the environment as a result decrease the content compounds of chrome (III) in waste solutions in 2.0 times, sulphates in 1.5 times, chlorides in 1.2 times, index of BOD in 2.0 times (table 4) and significantly shorten duration of tanning in 5.0 times (table 3).

Table 4. Description of waste tanning waters.

Index		Exhaustion of tanning agent (%)	BOD (g O_2/l)	COD (g O_2/l)	Content in the waste solutions (g/l)		
					chrome oxide	sulphates	chlorides
Technology	new	71.9	2.3	7.0	3.7	6.8	16.7
	known	54.4	4.7	8.2	7.3	10.1	19.4

4. Conclusions

To improve the technology of leather production the possibility of using a polymeric material based on acrylic acid to processing the bated pelt of sheep skin before chrome tanning and replace the pickling was investigated. It was found experimental the increasing of penetration and fixation of chrome tanning agent in structure of derma.

Using mathematical planning of the experiment, the optimal processing parameters were determined: consumption of polymeric material 1.0%; chrome tanning agent 1.6%; duration of polymeric processing 1.5h.

New pickling-free chrome tanning technology of garment leather production promotes the intensification of tanning in 5 times, increases the thermostability of leather on 13.5 °C, yield of thickness to 22.1%, exhaustion of tanning agent to 17.5%.

This improves the composition of waste solutions: index of BOD decreases in 2.0 times, content of chrome oxide in 1.9 times, sulphates in 1.5 times, chlorides in 1.2 times.

References

[1] Mihaylov A N 1953 *Chemistry tannin agents and tanning process* (Moscow: Legprom)
[2] Saravanabhavan S, Rao J, Nair B 2005 Integration of chrome tanning and wet finishing process for making garment leathers *Journal of the American Leather Chemists Association* vol 100 pp 225-232
[3] Shalbuev D M 2006 *Practical work on evaluation of the quality of waste water in the leather and fur enterprises* (Ulan-Ude: ESSTU)
[4] Pustylnik Y I 2008 Different types of tanning leather *Leather and Shoes* (St. Petersburg: Legprombissnes) chapter 3 pp 22–24
[5] Maistrenko L A, Andreyeva O A 2013 Developing of resource-saving technology of chrome tonnage for garment sheep leather *Materials of the III International research and practice conference Science, Technology and Higher Education* (Canada, Westwood) vol 2 pp 454-456
[6] Lukianets L, Andreyeva O 2009 Investigation of characteristics of new polymeric materials for leather and fur production *Visnyk KhNU (Message 1)* vol 4 pp 208-211
[7] Danilkovich A G 2006 *Laboratory practice at the rate of chemistry and technology of leather and fur* (Kyiv: Phoenix)
[8] Balberova N, Mikhaylov A N *et al* 1986 *Directory of tanner (Technology)* (Moscow: Legprombytizdat)
[9] Pinchuk S I 2008 *Organization of experiment in modelling and optimization of technical systems* (Independent Publishing organization, Diva)
[10] Maxfield B 2009 *Essential MathCAD for Engineering Science and Math* 2nd edition (Academic Press, Science and Math)
[11] Gorbachev A A *et al* 2007 *Basis for the creation of modern production technologies* (Kyiv: KNUTD)

IOP Conf. Series: Materials Science and Engineering **111** (2016) 012025 doi:10.1088/1757-899X/111/1/012025

Green Alternative binders for high-voltage electrochemical capacitors

V. Khomenko[1], V. Barsukov[1], O. Chernysh[1], I. Makyeyeva[1], S. Isikli[2] and F. Gauthy[2]

[1]Kiev National University of Technologies & Design, 2 Nemirovich-Danchenko str., Kiev 02011, Ukraine

[2]Solvay S.A. Functional Nanomaterials, Rue de Ransbeek 310, B-1120, Brussels, Belgium

E-mail: v.khomenko@i.ua

Abstract. Different polymer binders have been studied for applications in positive and negative electrodes of lithium-ion capacitor (LIC). The current work describes also some new approaches for changing the electrode fabrication process from organic to aqueous solvents in order to reduce the cost of the electrode making process. LIC electrodes made from Solef® PVDF aqueous dispersions have revealed a superior electrode quality and a high electrochemical performance. These achievements have demonstrated the possibility to develop an environmentally friendly production process of LIC.

1. Introduction

An electrochemical device would best fit for the automotive technology once the compromise between specific energy and power density is established. Even though Li-ion batteries (LIB) are famous storage technologies representing the highest energy densities, and can now be designed to deliver also a high power, some important drawbacks such as limited cycle life and safety problems give rise to important concerns for their widespread use, particularly in transport area. On the other hand, carbon-carbon electric double layer capacitors (EDLC) have relatively unlimited cycle-life, very high specific power but not enough energy densities. A new type of nonaqueous hybrid device, also known as lithium ion capacitors (LIC), was first introduced by Amatucci et al.[1] This hybrid system basically combines a battery negative electrode and an EDLC type positive electrode to profit the advantages of both technologies [2].

Despite several factors impacting LIC power and energy densities, electrode manufacturing parameters merit special attention to reach the best performance. In this sense selection of binder is one of the crucial steps of electrode manufacturing. A high performance binder should establish a high adhesion of the electrode active materials to the current collector along with the ability to form a good electric network between the active material and conductive carbon in order to improve both the electron transport and the diffusion of the lithium ion. Optimization of the different binder systems for electrode fabrication is one of the ways to develop high performance electrochemical storage devices.

IOP Conf. Series: Materials Science and Engineering **111** (2016) 012025 doi:10.1088/1757-899X/111/1/012025

Among different polymer binders, poly(vinylidene fluoride) (PVDF) have been widely applied as binder due to their excellent chemical, thermal and electrochemical stability [3]. Nevertheless, their main shortcoming is the organic solvent which is generally used to achieve the complete dissolution of PVDF in order to provide a good adhesion. Currently, the common organic solvent for preparation of the electrode slurry is N-methyl-2-pyrrolidone (NMP).

Recent approaches have been devoted to avoid hazardous organic solvents and to replace them with more environmentally friendly alternatives. Aqueous-based solutions have been proposed to develop a greener process for electrode manufacturing. The use of organic solvents is avoided and replaced by water as solvent media. A lot of efforts had been put into practice to search for alternative water-based binders and to improve the electrochemical performance. Carboxymethyl cellulose (CMC) [4,5], polyacrylic acid (PAA) [6,7] and polyvinyl alcohol (PVA) [8,9] have been extensively tested. The system based on Styrene-Butadiene Rubber (SBR) and CMC is the most studied water-based binary binder combination.

In this study a pioneering slurry concept targeting the use of an environmentally benign slurry solvent and aqueous-based binder systems were investigated for constructing LIC electrodes. For this purpose Solef® PVDF aqueous dispersions were successfully employed as novel binders for LIC's electrodes. Both adhesion and electrochemical performance evaluation in coin cells were realized and compared to NMP-based electrode development. As a consequence of the utmost efforts to change the binder type from organic to aqueous. Solef® PVDF aqueous dispersions alternatives were employed as binders due to desired electrochemical and chemical stability. LIC electrodes are manufactured, and complete electrochemical tests were performed in comparison to the well-established CMC-SBR system.

2. Experimental

In this work, versatile electro-chemical testing has been conducted with different active materials in order to estimate their possible application for lithium-ion capacitors. Electrodes were composed of 80-91% of active materials (graphite or activated carbon) type, 5-10% of the binder, and 1-5% of carbon percolators. In current work activated carbon YP 50-F from Kuraray Co. selected for the positive electrodes of LIC. SLK1520P graphite sample from Superior Graphite Co had been selected for the positive electrodes of LIC. The carbon C65 from Timcal was used as a percolator for negative and positive electrodes. Slurries for electrode casting were prepared from a mixture of the active materials and solutions of binder. Tape casting with 100 μm gap was carried out by "Doctor blade" method. Dry electrodes, with a thickness within the range of 20-80 μm, were calendared by Hot Rolling Machine from MTI. Then electrodes were cut by the MTI disk cutter into 15 mm circular samples. Later the samples further dried at 120°C and transferred to the argon filled MBraun glove box for storage until cell assembling took place. The electrochemical tests carried out using a coin-type cell (CR-2025 from MTI). The electrochemical performance of materials is examined using different measurement techniques. Electrochemical investigations were performed using the multi-channel potentiostat/galvanostat VMP3 from Princeton Applied Research (UK).

3. Results and Discussion

The composite electrodes need to have enough mechanical stability, including good adhesion of the active materials to the substrate and high adhesion between the particles of active materials to get the best electrochemical performance. This is because much attention dedicated to explore the material adhesion to the current collector by varying electrode's composition and binders and electrode processing. Adhesion has been measured by peeling test following ASTM D 3359. In the current work electrodes of LIC are prepared from slurries based on activated carbon and different binders. Also, we tested electrodes made from slurries containing different solvents such as NMP or water. In this investigation, we clarified affects which influenced on the adhesion of active materials.

IOP Conf. Series: Materials Science and Engineering **111** (2016) 012025 doi:10.1088/1757-899X/111/1/012025

The table 1 shows the summarized results of adhesion tests applied on electrodes made according to different conditions. The results lead to a conclusion that applying the proper processing conditions, electrodes which made using water based binder show the same adhesion level electrodes based on PVDF polymer processed in NMP. Thus, replacing of NMP by water-based binders is a major step towards realizing an environmentally friendly, cost efficient and safer method of electrode development.

Table 1. The summary results of adhesion tests for electrodes with different binder materials

Composition of electrodes (%, w/w)	Electrode surface of cross-cut area from which flacking has occurred		Conclusion
	After incisions	After removing tape	
Activated Carbon (YP-50F)-90 %, C65-5%, PVDF (NMP)-5%	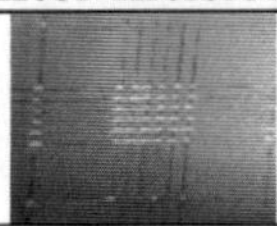		Bad adhesion
Activated Carbon (YP-50F)-85 %, C65-5%, PVDF (NMP)-10%	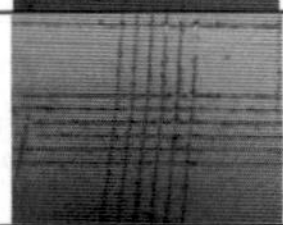		Good adhesion
Activated Carbon (YP-50F)-89 %, C65-5%, CMC-2%, SBR-4%	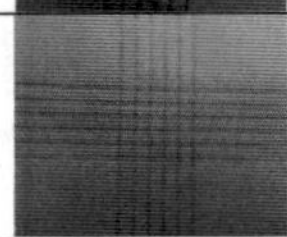		Good adhesion
Activated Carbon (YP-50F)-89 %, C65-5%, CMC-2%, PVDF Latex-4%	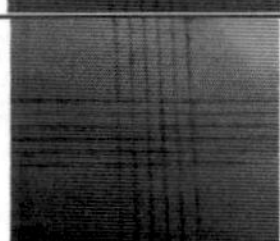		Good adhesion
Graphite (SLK1520P)-92%, C65-1%, PVDF (NMP)-7	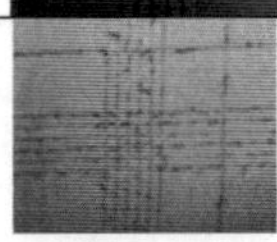		Good adhesion
Graphite (SLK1520P)-93%, C65-1%, CMC-3%, PVDF Latex-3 %	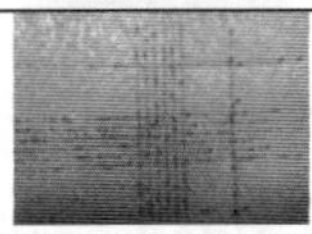	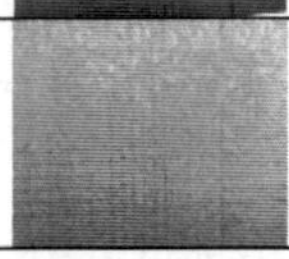	Good adhesion

3.1. Electrochemical performance of positive electrode

The internal resistance is the key parameter for determining power, energy efficiency of a supercapacitor. In this work the internal resistance of supercapacitor has been determined by Electrochemical Impedance Spectroscopy (EIS) techniques. As Fig.1 shows the internal resistance of supercapacitor can be easily obtained from Nyquist plot. The internal resistance can be specified as Equivalent Series Resistance (ESR) and Equivalent Distributed Resistance (EDR). ESR is the resistance due to all the resistive components within the supercapacitor. The EDR resistance includes the ESR resistance and an additional contribution from the charge redistribution process in the electrode.

In this work, water-based binders were employed as an alternative slurry solvent instead of NMP. Fig. 2 shows Nyquist plots of the EDLC cells assembled with activated carbon (AC) electrodes based on different binders. Fig. 10 shows Nyquist plots of the EDLC cells assembled with AC electrodes based on different water based binder. Comparing the results in table 2 leads to the conclusion that AC electrodes, made from a water-based binder, show better electrochemical performance as standard

PVDF processed in NMP. Thus, according to our results, the proposed techniques should be potential alternatives to a common fabrication process of lithium-ion capacitor electrodes based on NMP.

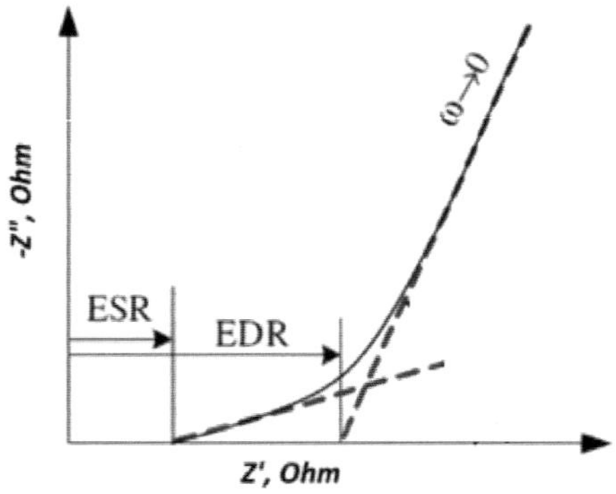

Figure 1. Determination of ESR and EDR from Nyquist plot

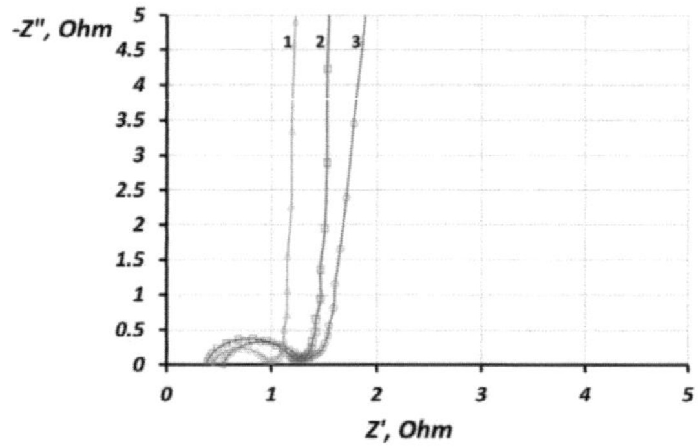

Figure. 2 Nyquist plots of the EDLC cells assembled with AC electrodes based on different water based binder:
1- ***CMC-PVDF Latex***
2- ***PVDF processed in NMP***
3- ***CMC-SBR***

Table 2. The summary results of electrochemical tests for EDLC capacitor based on electrodes with different binder materials

Composition of electrodes (%, w/w)	***R_{ESR}, Ohm***	***R_{EDR}, Ohm***
YP-50F-89 %, C65-5%, CMC-2, SBR-4%	0.51	1.53
YP-50F-89 %, C65-5%, CMC-2%, PVDF latex -4 %	0.44	1.15
YP-50F-89 %, PVDF (NMP)	0.71	1.45

3.2. Electrochemical performance of negative electrode

The electrochemical investigation of all anodes was started with the first lithiation performed against metallic lithium. Fig. 3 shows the initial charge/discharge curves of the SLK1520P graphite electrode which was prepared using the standard method with NMP as solvent for electrode slurry. The table 3 sums up initial charge/discharge performance for the SLK1520P graphite samples from Superior Graphite Co. The difference between the specific charge of the first lithiation, and the reversible specific charge (*Qrev*) of the following de-lithiation is defined as the irreversible capacity loss of the graphite electrode. The irreversible capacity loss (*Qirr*) is a measure of the capacity lost due to SEI formation on graphite anode. The reversible capacity of SLC1520P graphite at the first cycle was of Q ~ 358 mA·h/g; this is quite close to the theoretical capacity. Irreversible loss of capacity at the first cycle came to 18 mA·h/g ($\Delta \sim 5$ %). The above-mentioned grade of graphite has been selected for LIC due to their lowest irreversible capacity. The same test was done for electrodes with different binders. Results of investigation are included in Table 3 as well. It seems that electrodes made with aqueous binders show the same electrochemical performance as standard PVDF processed in NMP.

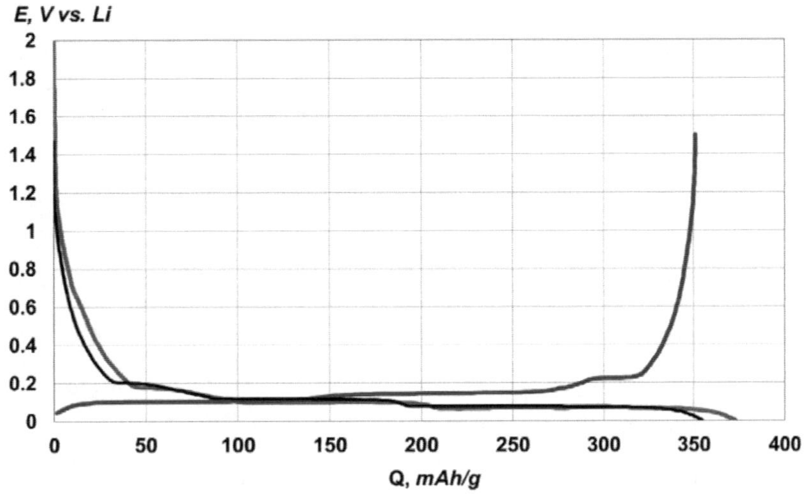

Figure 3. Charge and discharge profile of the half - element based on the reference SLC1520P graphite at the current density C/10.

Table 3. Electrochemical performance of anode based on different binder materials

Anode composition	*Qrev, mA·h/g*	*Qirr, mA·h/g*
SLK1520P -91%, C65-1%, PVDF(NMP)-8%	358	18
SLK1520P -93%, C65-1%, CMC-3%, PVDF Latex-3%	362	17
SLK1520P -93%, C65-1%, CMC-3%, SBR-3%	356	23

The electrodes based on aqueous binders are very promising in terms of electrochemical performances of the negative electrode. In combination with the advantage of enabling usage of water as solvent for the electrode preparation process, these electrodes certainly deserve further investigation and optimization efforts.

3.3. Electrochemical performance of lithium-ion capacitor

The electrochemical performances of the lithium ion capacitors were investigated. LICs were assembled with pre-lithiated graphite anode and AC cathode based on aqueous binder. Fig. 3 and 4 show the charge-discharge curves of LIC based on water-based binder with a potential range of 2.0-4.5 V at a current density of 1C.

The Ragone plots are illustrated in Fig. 4, which were obtained by calculating the energy density and power density at different current densities. It should be noted that both the energy density and power density of LIC based on aqueous binder (CMC- PVDF latex) improved in comparison to LIC with electrodes based on PVDF processed in NMP. Designed LIC with PVDDF latex binder has the maximum energy density of 80 W·h/kg. Also the highest power density (of about 20 k·W/kg) is possible to achieve. Thus advanced LIC show the best combination of energy density and power density, what is especially important for automotive application.

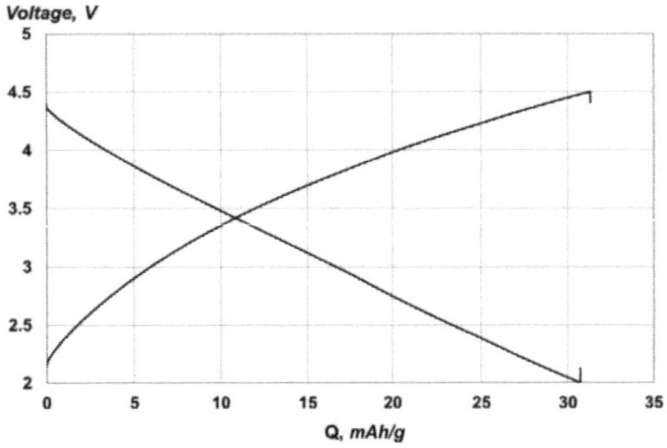

Figure 3. Charge-discharge curves of the LIC assembled with activated Carbon (cathode) and Graphite (anode). Specific capacity calculated per mass of cathode.

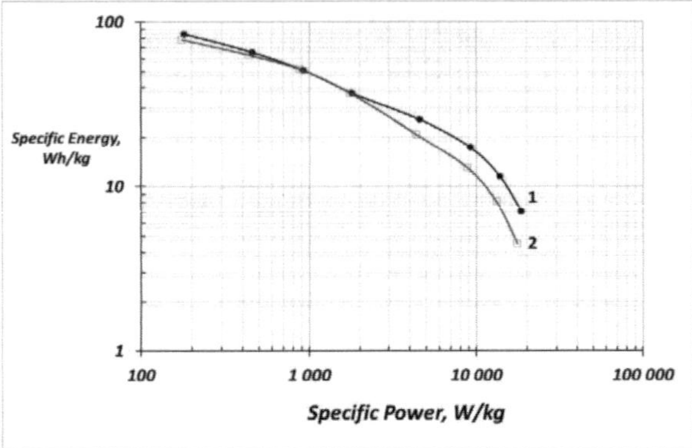

Figure 4. The Ragone plots for the LIC accepted with electrodes using different types of binders:
1-CMC-PVDF Latex
2-PVDF processed in NMP

4. Conclusion

The obtained results enlighten the promising prospect of innovative slurry concept for the fabrication of LIC electrodes. These achievements have demonstrated the possibility to develop LIC environmentally friendly production processes. Applying the proper processing conditions, electrodes made with water-based binders (CMC/SBR, CMC-PVDF Latex) show the similar electrochemical performance as standard PVDF processed in NMP. Through comparing the electrochemical performance different electrodes based on SBR and novel Solvay product such as PVDF latex it can be concluded that PVDF-latex shows better performance than SBR.

Acknowledgement

Authors acknowledge European Commission for the financial support of these researches in the framework of the FP7 "Energy Caps" project. We would like also to thank colleagues from Solvay R&I Center, Bollate, Italy for providing the advanced binders (Solef® PVDF water-based dispersions).

References

[1] US Patent 6,252,762.
[2] Amatucci G, Badway F, Du Pasquier A, Zheng T 2001 *Journal of Electrochemical Society* **148** A930
[3] Amin-Sanayei R and Heinze R 2009 *NASA Aerospace Battery Workshop*, Huntsville, AL
[4] Porcher W, Lestriez B, Jouanneau S and Guyomard D 2009 *Journal of the Electrochemical Society*, **156**, A133
[5] Wang Z, Dupre N, Gaillot A-C, Lestriez B, Martin J-F, Daniel L, Patoux S, and Guyomard D 2012 *Electrochim. Acta*, **62**, 77.
[6] Zhang Z, Zeng T, Qu C, Lu H, Jia M, Lai Y and Li J 2012 *Electrochim. Acta*, **80**, 440–444
[7] Zhang Z, Bao W, Lu H, Jia M, Xie K, Lai Y and Li J, 2012, 1, 2012–2015.
[8] Chaouat M, Visage C L, Baille W E, Escoubet B, Chaubet F, Mateescu M A and Letourneur D 2008 *Advanced Functional Materials*, **18**, 2855.
[9] Park H K, Kong B S and Oh E S 2011 *Electrochem. Commun.*, **13**, 1051.

IOP Conf. Series: Materials Science and Engineering **111** (2016) 012026 doi:10.1088/1757-899X/111/1/012026

Composite materials for protection against electromagnetic microwave radiation

I V Senyk, V Z Barsukov, B M Savchenko, K L Shevchenko, V P Plavan, Yu V Shpak and O A Kruykova

Kyiv National University of Technologies and Design Nemirovich-Danchenko street, 2 Kyiv 01011, Ukraine

E-mail: v-barsukov@i.ua

Abstract
A fairly wide range of carbon-polymer composite materials was synthesized and studied in terms of their potential to protect people and electronic equipment from exposure to electromagnetic radiation (EMR). The materials studied included three main groups: (1) PVC polymer composites filled with various carbon-containing fillers (colloidal graphite, thermally expanded graphite, acetylene black, graphitized carbon black, carbon nanotubes, graphene) at concentrations ranging from 5 to 20%; (2) carbon cloth - commercial and modified with nano-metal additives (e.g., nanoparticles of Cu, TiN, etc.); (3) highly-filled polymer-carbon composites in the form of paint. The transmission rate α of electromagnetic radiation was investigated for such materials in the frequency range of 10 GHz as well as their electrical conductivity.
The results showed that the shielding ability of the materials of group (2) is significantly higher than that of the materials of group (1), which is probably due to the presence of strong internal skeleton of conductivity. Nevertheless, some highly-filled mixed polymer-carbon composites in the form of paint demonstrate even more shielding ability than carbon cloth and could be used for the defence against EMR.

1. Introduction

Electromagnetic radiation (EMR) influences population and electronic equipment because of increasingly widespread use of television and radiobroadcasting, microwave emitting devices and other modern technologies. However, it is not the entire spectrum of the signal and the higher harmonic components that is harmful to living things, but the waves whose lengths fall in the millimetre and centimetre range. Circulatory system, brain, eyes, immune and reproductive systems are known to be most susceptible to electromagnetic fields. It encourages the search for materials and technologies that can protect humans and equipment from ultra-high frequency (UHF) electromagnetic radiation, electromagnetic terrorism and weapon of directed energy [1, 2].

Some investigations of electromagnetic losses in carbon-epoxy composites have shown the sufficient influence of the type of carbon filler (multi-walled carbon nanotubes, carbon foams, etc.) and its content on the value of losses [3-5].

The main objective of this work was to create electroconductive polymer systems based on inexpensive and easy available carbon agents and polymer matrixes, which could block EMR and ensure quite good electromagnetic interference shielding effectiveness [6].

2. Experimental

A fairly wide range of composite materials have been synthesized and studied in terms of their potential to protect people from exposure to electromagnetic radiation.

The main materials studied included three main groups:
(1) polymer composites filled with various easy available carbon-containing fillers (colloidal graphite, thermally expanded graphite, acetylene black, graphitized carbon black) at concentrations ranging

from 5 to 20%;
(2) carbon cloth – commercial as well as modified with nano-metal additives (e.g., nanoparticles of Cu, TiN);
(3) highly-filled polymer-carbon composites in the form of paint.

2.1. The methodology for preparing conductive composites based on carbon fillers

In general, the methodology for preparing conductive composites involves mixing the components (polymer, softener, carbon fillers, etc.) and their subsequent plasticization at high temperature. We used the following specific methods:

In the first case the polymer was diluted in dioctyl phthalate (CAS #117-81-7) as a softener; after this the carbon filler (5-20 %) was added to the product and mixed for 1 hour. The finished composite was sputtered on the substrate and dried for 3 min at 200 ^{0}C.

The second method for preparing composites also included dilution of the polymer-powder in the softener, but the carbon filler was previously mixed with wetting agents (soybean oil, ethanol, etc.) and after that was mixed with the product. The conditions of drying were the same as in the first case.

The third method was similar to the second one, but the composites contained 60-80% of carbon additives. As a result, carbon highly-filled polymer paint was formed and then mapped on the surface of polymer sample.

2.2. The methodology for modifying carbon cloth with nanoparticles

The carbon cloth sample was placed in a container so that it would lie entirely in one plane. After this the sample was wetted with 2 ml «basic» $CuSO_4$ solution until complete wetting. Then, there was prepared the following mixture of substances: 2 ml of KI and 2 ml of $Na_2S_2O_3$ in a clear flask. The resulting mixture was poured onto a previously prepared re-wetting sample. As a result of the chemical reaction, CuI particles formed directly on the carbon fibres and between the fibre spaces. The sample was dried in air at temperatures of 50-100 ^{0}C to constant weight. Also the carbon cloth sample modified with TiN was produced by vacuum deposition techniques.

2.3. The measurement technique

Electrical conductivity and resistance are important parameters in determining the shielding effectiveness that the EMI screen or other shields would provide. They are much more critical at higher frequencies, where surface resistivity is predominant. Surface resistivity and volume resistivity are commonly tested for characterizing electrical conductivity of EMI screening materials as a quick and practical quality check.

2.3.1. The measurements of specific volume resistance

The measurements of specific volume resistance were carried out initially according to the common "three-electrode" method (technique 1) and then using the "four-electrode" method (technique 2).

Technique 1 involves measuring the current passing through the sample at a constant voltage. In this technique, constant voltage from the voltage source is applied to two high-voltage electrodes. A current is going through the thickness of the sample to the working electrodes or through the surface to the protective electrode (PE). From the PE the current is directed to the ground by passing the working electrodes.

Technique 2 uses the following four electrodes [2]:

1) two power current electrodes on the ends of the sample;
2) two measuring "potentiometric" electrodes in the middle part of the sample.

The voltage source was connected via the ammeter to the current electrodes and digital voltmeter was connected to the "potentiometric" electrodes.

The main condition of measurement accuracy is that the input impedance of digital voltmeter should be much higher than the possible contact resistance.

This technique enables one to exclude the contact resistance and to measure the true resistance of composite layer due to the negligible currents between measuring electrodes. As our experiments have shown, the contact resistance in our samples was quite high. So, technique 2 is more preferable in our case.

The volume electrical resistivity, ρ, is expressed as the electrical resistance of a material per unit length multiplied by the cross-section area:

$$p_v = U \cdot a \cdot b / l \cdot I \ (1) \quad ,$$

where U is the voltage, V; I is the current, A; a is the thickness, cm; b is the width, cm; l is the length between the measuring electrodes, cm.

2.3.2. *The measurements of EMR blocking*

Evaluation of the absorption of electromagnetic energy by the sample of composite, which was fixed between the horn antennas (figure 1), was carried out by comparing the measured values of the reflection coefficient module and transfer coefficient. Electromagnetic properties of radio absorption coatings were studied in the frequency range of 10 GHz.

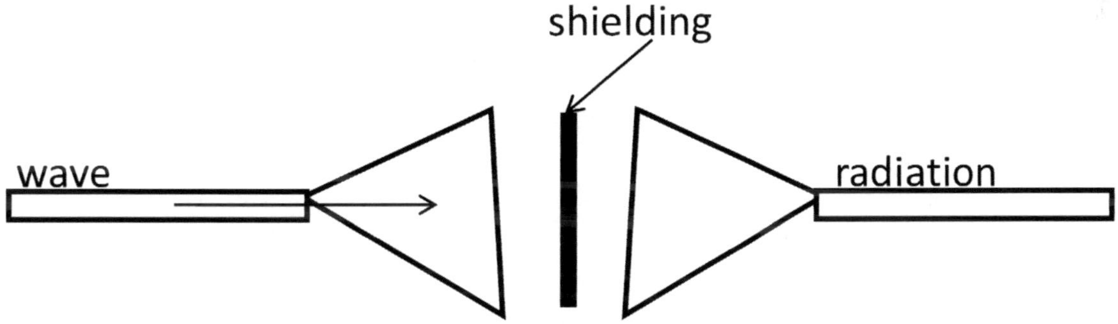

Figure 1. Simple scheme of the shielding measurement

3. Results and discussion

Fig. 2 shows the dependence of the resistance on the nature and concentration of carbon filler for the more typical carbon materials, which were investigated using technique 2.

Graphitized carbon black (commercial grade PUREBLACK ® Carbons) and thermally exfoliated graphite (commercial grade ABG 1010) from Superior Graphite Co. Chicago, IL, USA have demonstrated the best conducting properties as conducting filler for the above-mentioned composites.

This effect can be explained not only by the well-developed surface area of these materials, which can reach and sometimes exceed the value of 50-65 m^2/g, but also by specific nano-structure peculiarities of such carbon materials. A high conductivity requires so called "high structure" materials, whose macromolecules have multiple "branches" consisting of individual carbon nano-particles [7]. These highly developed nano-sized branches are believed to create additional contact points, thus resulting in efficient conductivity enhancement at low concentrations of carbon additives. As a result, such materials are usually more conductive than traditional forms of acetylene-type conductive carbon black.

The percolation threshold for the graphitized carbon black and thermally exfoliated graphite is less than 5%. It is not possible to fix this threshold from Fig. 1 as we selected the range of carbon concentration from 5 to 20%.

At the same time, Ukrainian Zaval'e colloidal graphite preparation (grade S-1) from the Zavalyevsky graphite plant [8] has demonstrated a classical dependence of resistance on the

IOP Conf. Series: Materials Science and Engineering **111** (2016) 012026 doi:10.1088/1757-899X/111/1/012026

concentration of conductive additive. The corresponding graph shows the clear formation of percolation threshold, which occurs at concentrations of carbon material ca 10-12% in polymer-graphite composite (figure 2).

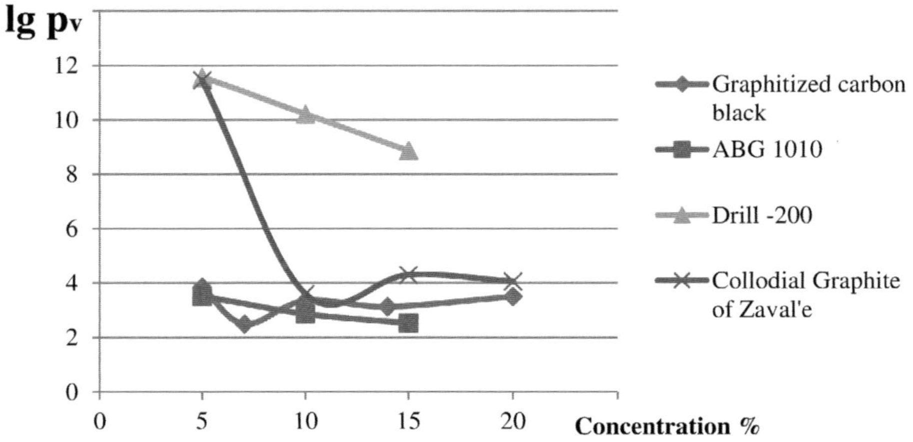

Figure 2. The experimental curves of dependence of the resistance vs the filler nature and concentration

The task of shielding electromagnetic waves is relevant both in the organization of EMC and in the protection of electronic equipment against powerful electromagnetic pulses. The dependence of shielding properties of materials on electrical conductivity is directly proportional (figure 3). We investigated the transmission rate, α, of electromagnetic radiation for such materials in the frequency range of 10 GHz (figure 3).

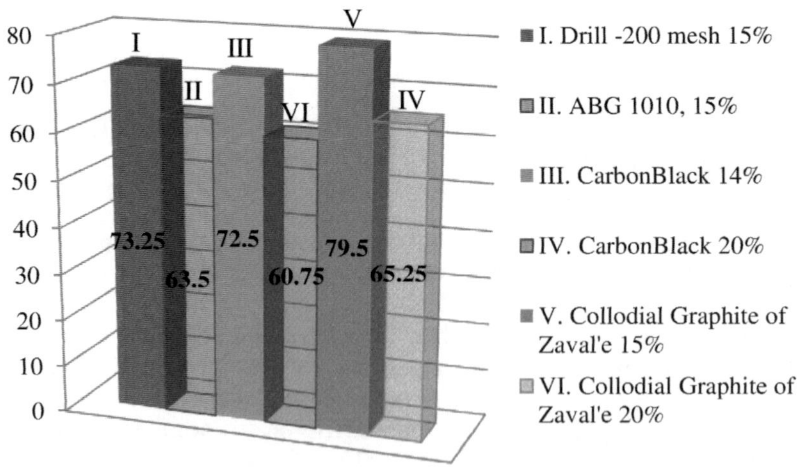

Figure 3. Value of transmission coefficient α of electromagnetic wave by composite sample, %

Fig. 3 shows that the highest effectiveness has been received for sample IV with 20% graphitized carbon black. The samples based on thermally exfoliated graphite and colloidal graphite have almost the same values, but these values of the transmission of electromagnetic energy are not enough for effective use. To create more efficient screening coverage, it is necessary to ensure maximum electrical conductivity of the sample. To fulfil this task, a new type of samples was developed, in particular sheeting paint with a graphite filler content of more than 60%.

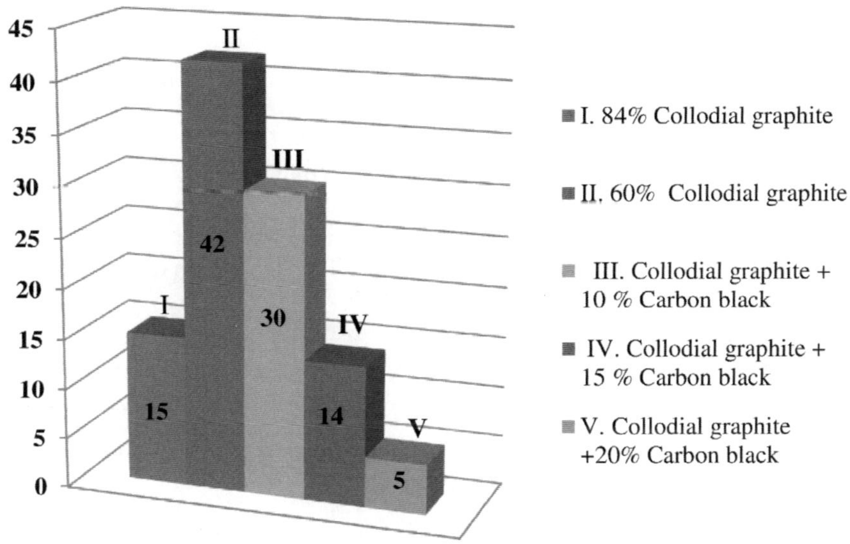

Figure 4. Value of transmission coefficient α of electromagnetic wave by composite sample, %

Figure 4 shows the coefficient of electromagnetic radiation transmission α of different composite materials. Colloidal graphite shows good characteristics, and it does not require additional operations (sifting, grinding, pre-wetting, etc.) in producing composite materials. Besides, it is based on home industry product, which makes it available [8].

The outside surface is an important attribute of the paint coating with a special function. It is important for the coating to have good adhesion and to be intact during its life. The high content of carbon fillers $\geq$ 80%, like that in colloidal graphite, provides decent protection from electromagnetic radiation, but leads to cracking of the system and, therefore, to delamination of the coating and poor performance. Adding even a small amount (10-20%) of graphitized carbon black can improve the composite system. Such a composite material is effective in terms of shielding properties and technological in terms of its ability to form a quality- surface uniform film.

Thus, a promising and competitive shielding composite material was created, which is capable of protecting important facilities and people from electromagnetic radiation and can ensure electromagnetic interference shielding effectiveness at least 10 dB in the range of about 10 GHz.

4. Conclusions

According to the obtained data it is clear that the shielding properties of the materials correlate in general with electrical conductivity (figures 3, 4).

It was shown that even a small addition of the material with a well-developed surface area (graphitized carbon black) improves effectiveness of electromagnetic wave absorption by graphite composite samples (figure 5).

The results showed that the shielding ability of the materials of group (2) is significantly higher than that of the materials of group (1), which is probably due to the presence of strong internal skeleton of conductivity. Nevertheless, some highly-filled mixed polymer-carbon composites in the form of paint demonstrate even more shielding ability than carbon cloth, could be used for effective defence of people and equipment from EMR and ensure electromagnetic interference shielding effectiveness at least 10 dB in the range ca 10 GHz.

IOP Conf. Series: Materials Science and Engineering **111** (2016) 012026 doi:10.1088/1757-899X/111/1/012026

References

[1] Heald M A, Marion J B 2012 *Classical electromagnetic radiation* (Dover Books on Physics, NY, USA)

[2] Celozzi S, Araneo R, Lovat G 2008 *Electromagnetic Shielding* (J. Wiley & Sons Inc., Hoboken, New Jersey, USA)

[3] Matzui L, Vovchenko L, Prylutskyy Yu, Korotash I, Matzui V, Eklund P, Ritter U, Scharff P 2007 Mat. Sci. Eng. C-Bio. S., **27**, 1007

[4] Macutkevic J, Kuzhir P P, Paddubskaya A G, Banys J, Maksimenko S A, Stefanutti E, Micciulla F, Bellucci S 2013 *J. Nanophotonics*, **7**(1) doi: 10.1117/1.JNP.7.073593

[5] Kuzhir P P, Paddubskaya A G, Shuba M V, Maksimenko S A et al. 2012 *J. Nanophotonics*, **6**(1) doi: 10.1117/1.JNP.6.061715

[6] Tong X C 2009 *Advanced Materials and Design for Electromagnetic Interference Shielding* (CRC Press Taylor & Francis Group, Boca Raton, FL, USA)

[7] Barsukov I V, Gallego M A, Doninger J E 2006 *J. Power Sources*, **153**(2), 288

[8] Zavalyevskiy graphite. Graphite grades (2015), Ukraine.
Web-site: http://zvgraphit.com.ua/en/chapter/view/1739/

Polymer sand composites based on the mixed and heavily contaminated thermoplastic waste

I Slieptsova, B Savchenko, N Sova, A. Slieptsov

Department of Polymer and Chemical Fibre Technology, Kyiv National University of Technology and Design, Nemirovich-Danchenko Street, 2, Kyiv 01011, Ukraine

E-mail: inna.slieptsova@yandex.ru

Abstract. This paper represents production and characterization of highly filled polymer composites based on recycled plastics and sand as a filler. Composites produced on the bases of polyolefin (polyethylene PE and linear low-density polyethylene LLDPE) and polyester (PET) were compared. Different modification techniques, including filler surface treatment, compatibilizer introduction and polymer blend formation were used. Influence of the composition on the composite mechanical properties was evaluated. Newly developed composite based on PET – polycarbonate (PC) blend demonstrates improved mechanical properties and higher thermal stability compared to polyolefin based composites.

1. Introduction

One of the most adverse result of human activities is formation waste plastic. More than 400 different kinds of industrial and domestic plastic wastes are known and many of those are identified as mixed plastic waste, which is not recycling friendly [1]. Recycling of mixed plastic waste is very important topic with great environmental and industrial influence. Today recycling options for mixed plastic are incineration, landfilling and mechanical recycling from which mechanical recycling is sustainable and can be economically profitable. A challenge for recycling technology is highly contaminated and mixed plastic waste. Finding recycling option for this type of waste is dramatically important task.

The main part of the heavily contaminated waste consists of such as polymers: polyolefins (PE&PP), polyesters (PET&PET-G), chlorinated polymers, like polyvinylchloride (PVC), and styrenics (PS, ABS). Mixed polyolefin waste can be used for formation of composites, including polymer-wood and highly filled composites. For mixed and highly contaminated polyesters recycling options include chemical depolymerization and advanced sorting technics. Development of simple mechanical process for such type of raw material is seen to have great industrial importance. Highly filled polymer composites based on mixed waste plastics and sand called as "polymer sand composites" normally have low mechanical and thermal properties, which significantly limits their application [2-3]. The aim of current study is to develop highly filled composites based on polymer binder in combination with different modification techniques.

IOP Conf. Series: Materials Science and Engineering **111** (2016) 012027 doi:10.1088/1757-899X/111/1/012027

2. Experimental

This study deals with highly filled polymer composites based on mixed mixed polymer waste. Following types of raw materials were used: polyethylene "Polimir" ™ PE-15803-020 (MFI=2-3), linear low density polyethylene "ExxonMobil ™ LLDPE LL4004EL, industrial grade recycled PE waste based on stretch film (R-LLDPE) (LLDPE modified with tactile additive) (MFI=3-4), virgin PET type 8200 (IV=0,8), industrial grade recycled PET (R-PET) (IV=0,7), extrusion grade PET glycol, PET-G, Skygreen S2008, polycarbonate TRIREX 3020U (PC), maleic anhydride grafted polyethylene copolymer (PEgMA), which was synthesized in the laboratory, functional silan Dynasylan® AMEO from "Evonik" ™. As filler, purified and dried sand was used after 75 mesh sieve classification.

2.1. Material preparation and testing

Before composite preparation all polymers were grinded to powder with average particle size 200 – 300 micron to facilitate homogenous mixing with the filler. Polymer components were dried in hot air oven at following conditions: PET – 8 hours at 150°C, PC and PET-G – 6 hours at 100°C, PE-g-MA – 2 hours at 60°C. Sand with a particle size from 0.2 mm to 1.2 mm was dried until constant weight before use. Silane treatment of sand was carried out in drum mixer by 5% aqueous solution of silane followed by drying in air oven. PE-g-MA copolymer was obtained by reactive extrusion in 22 mm laboratory twin-screw extruder. Powdered PE was pre-mixed with 2.0 weight % of initiator – dicumyl peroxide (DCP) in high speed mixer and 3.0% of maleic anhydride. The resulting ternary system was subjected to extrusion at 130-160°C to obtain pellets. Then the pellets were dried under vacuum to remove volatiles. Degree of grafting, measured by titration, was in range from 1.2 to 1.3% .

The polymer-sand composite materials were obtained in two ways. By the first way, the polymer powder and filler were mixed at room temperature and then, heated in muffle furnace at 250 - 270°C for 10 - 15 minutes and mixed in the heated twin rotor mixer. In the second case - polymers were melted in the single screw extruder and then fed in the twin blade mixer with hot sand by keeping the mixing time mixing time from 3 to 10 minutes. After mixing soft polymer-sand composite (PSC) was compression molded in hydraulic press.

Testing samples were mechanically milled or cut from bulk part produced by compression molding. The density (ρ) was determined according to GOST 15139. Bending strength (σb) was determined according to GOST 9550, while tensile elongation and strength were determined according to GOST 11262. Heat deflection temperature (HDT) was determined according to GOST 15088 (comparable to ISO 306). Melt flow index of composite material was determined at weight of 21.6 kg. For morphology characterization, optical microscope with digital camera was used.

3. Results and discussion

In the study properties of PSCs based on polyolefins and polyesters were compared. Initially, for comparison basic composite samples were made by using neat polyolefin as binder. Basic properties of PSCs based on polyolefin binder are listed in the Table 1.

With increase of filling degree from 50% to 80% decrease in tensile strength and increase in compressive and bending strength is observed. From technological point of view R-LLDPE based compositions have certain advantages because of higher flow and plasticity which results in easier manufacturing. Advantage of R-LLDPE based compositions is also greater flexibility. In general mechanical properties and HDT of polyolefin PSC are low which results in non-demanding application of such material. Current industrial practice shows that it is possible to produce large, heavy, parts thickness parts for non-demanding application like floor tiles, grooves and so.

Table 1. Properties of polyolefin PSCs

Polymer binder	Filling degree, mass %	Density (ρ), kg/м^3	Bending strength (σb), MPa	Tensile elongation, %	Tensile strength, (σt), MPa	Tensile modulus, Ep, MPa	MFI, g/10 min	HDT, °C
PE-158	50	1650	21	12.5	16.1	850	3.1	87
PE-158	65	1720	24	8.4	14.4	900	1.6	92
PE-158	80	1950	30	5.3	12.2	1300	1.2	97
LLDPE	50	1630	19	14.2	15.6	790	2.8	87
LLDPE	65	1745	21	9.5	15.2	900	1.9	91
LLDPE	80	1920	28	6.3	13.5	1150	1.5	95
R- LLDPE	50	1635	16	15.4	17.2	550	4.3	75
R- LLDPE	65	1768	20	11.3	16.4	750	3.1	84
R- LLDPE	80	1925	28	7.6	14.6	1050	2.1	86

Usage of PET or PET-G as polymer matrices provides a possibility to produce composites with improved mechanical properties and higher temperature resistance. Manufacturing of polyester PSCs was carried out at relatively higher temperature (260 – 270°C) resulting in enhanced flow of the melted composition. Because of PET crystallization process, which occurs during cooling of the molded part such PSCs are much more rigid and heat stable. For amorphous PET-G based compositions heat stability is much lower that for PET based systems. Table 2 shows properties of PSCs based on thermoplastic polyesters. In comparison to polyolefin type PSC, polyester PSCs show advanced mechanical properties and thermal stability. HDT and mechanical strength of PET and R-PET based composites are more than two times higher than for polyolefin PSC. Possible explanation of behavior is better wetting of filler particles with PET melt and specific strength of polymer matrix. Amorphous PET-G composites demonstrate similar mechanical properties but much lower HTD. PET and R-PET composites, however, are more brittle than PET-G based systems PET-G.

Table 2. Properties of polyester PSCs

Polymer binder	Filling degree, mass %	Density (ρ), kg/м^3	Bending strength (σb), MPa	Tensile elongation, %	Tensile strength, (σt), MPa	Tensile modulus, Ep, MPa	MFI, g/10 min	HDT, °C
PET	50	1840	74.2	4.4	31.3	1858	12.6	210
PET	65	2150	85.4	3.5	30.5	2250	8.3	215
PET	80	2350	105.2	2.1	24.6	2500	6.4	225
R-PET	50	1920	72.5	3.3	26.4	1950	14.5	216
R-PET	65	2205	86.4	2.2	24.3	2300	12.1	225
R-PET	80	2390	92.6	2.6	28.2	2650	13.0	220
PET-G	50	1730	76.4	8.3	31.5	1650	9.2	95
PET-G	65	1950	77.3	6.5	36.3	1800	6.3	98
PET-G	80	2150	87.4	5.7	34.7	1950	4.5	102

Further study was focused on investigation of influence of filler surface modification on PSC properties. Initially silane coupling agent was dispersed in water (5 mass% aqueous dispersion) and stored for 1 hour to promote hydrolysis. Then sand was treated with silane-water suspension (3 mass % from the sand content) and mixed for 30 min. in a tumble mixer. Finally silane treated sand was dried in an oven at 80°C. Properties of silane treated sand modified PSCs are shown in the table 3.

Table 3. Properties of surface treated PSC

Binder polymer	Filling degree, % mass	Density (ρ), kg/м^3	Bending strength (σb), MPa	Tensile elongation, %	Tensile strength, (σt), MPa	Tensile modulus, Ep, MPa	MFI, g/10 min	HDT, °C
R-PET	65	2160	89	5.4	26.5	2400	14.6	221
R-PET	80	2380	98	3.7	32.3	25000	16.5	220
PETG	65	1970	68	8.2	38.6	1700	7.8	97
PETG	80	2145	76	6.4	36.5	1890	7.5	100

Silane treated PSC show enhanced mechanical properties, which can be due to interfacial adhesion between filler and polymer matrix.

Another way of modification of PSCs is by enhancing polymer matrix adhesion to filler by using of compatibilizing agent. Graft copolymers of PE with maleic anhydride are common adhesion promoters between filler and polymer matrix. Compatibilizer was premixed with polymer matrix by using single screw extruder, then granulated and used for composite forming. Table 4 shows properties of polyester and polyolefin PSCs modified by graft copolymer. Modification lead to some properties enhancement like tensile strength and most, especially, elongation that increased simultaneously with compatibilizer loading level. The most visible property enhancement occurs in the case of polyolefin-based system, which can be associated with good interaction of PE158 with polyethylene backbone of the compatibilizer. Introduction of PEgMA also lead to reduction of melt viscosity and improvement of surface finish of the test specimens.

Table 4. Properties of PEgMA modified PSC

Polymer binder	Filling degree, mass %	Density (ρ),kg/м^3	Bending strength (σb), MPa	Tensile elongation, %	Tensile strength, (σt), MPa	Tensile modulus, Ep, MPa	MFI, g/10 min	HDT, °C
PET-G + PE-gMA-2%	65	1920	68.2	11.5	37.3	1650	8.4	101
PET-G + PE-gMA-2%	80	2095	79.4	9.5	35.4	1900	9.2	103
PET-G + PE-gMA-5%	65	1930	64.3	14.4	34.7	1700	9.2	101
PET-G + PE-gMA-5%	80	2080	80.5	11.3	34.3	1950	10.8	102
PE158 + PE-gMA-2%	65	1730	23.6	18.5	13.6	800	3.5	93
PE158 + PE-gMA-2%	80	1920	25.4	7.6	21.4	1350	2.3	98
PE158 + PE-gMA-5%	65	1705	23.3	21.3	14.5	900	4.8	93
PE158 + PE-gMA-5%	80	1960	24.5	11.2	15.6	1300	2.5	99

Another way of modification of PSCs is formation of matrix from polymer blends. For this purpose bisphenol polycarbonate (PC) was used, mainly, because of good technological compatibility with PET. In the chosen processing temperature range PC and PET have similar values of melt viscosities and both are readily available in post-consumer polymer waste stream. Properties of PSCs based on PC/PET blend are summarized in Table 5.

Introduction of PC into PET matrix leads to great mechanical property enhancement and allows stable production of composite with 90% filling level. For experimental reason recycled PC derived from CD\DVD discs was also used for modification of PSCs. Obtained results shown similar modification efficiency as virgin PC.

Table 5. Properties of PSCs based on PC/PET blend

Polymer binder	Filling degree, mass %	Density (ρ), kg/м^3	Bending strength (σb), MPa	Tensile elongation, %	Tensile strength, (σt), MPa	Tensile modulus, Ep, MPa	MFI, g/10 min	HDT, °C
PET +PC 5%	65	2130	92.3	4.3	39.1	2400	9.2	211
PET +PC 5%	80	2450	118.6	3.5	31.3	2700	5.6	220
PET +PC 5%	90	2750	91.2	1.2	21.5	3000	3.2	225
PET +PC 10%	65	2170	105.5	6.3	46.5	2500	9.5	201
PET +PC 10%	80	2380	123.4	4.7	33.7	2700	6.2	212
PET +R-PC 10%	80	2420	118.4	4.2	30.2	2750	7.3	214

4. Conclusion

Introduction of PC or RPC into PSCs leads to the greatest property enhancement over other modification techniques and can be simply realized in existed production flow. Recycled PC is readily available and is inexpensive, if the waste stream is contaminated, like CD\DVD discs waste. This modification technology of PSCs has great economic perspectives and will be target for future study. In the case of polyolefin PSCs properties enhancement occur if graft copolymer is added but economical application is possible only with own copolymer production.

Polymer sand composites based on contaminated polyester waste have much higher properties in comparison to the systems based on polyolefines. Application of such composites allows to produces thinner and lighter parts with high rigidity and excellent thermal stability. They claim attention for further detailed scientific and industrial investigation and implementation.

5. References

[1] Tawfikand M E and Eskander S B 2006 Polymer concrete from marble wastes and recycled poly(ethylene terephthalate) *Jr. of Elastomers and Plastics* **38** pp 65–79

[2] Saikia N J and de Brito 2013 Waste polyethylene terephtalate as anaggregate in concrete *Materials Research* **16** pp. 341–350

[3] Pogrebnyak (Slieptsova) I L, Sova N V, Savchenko B M, Pakharenko V A, Moisyuk V S 2013 A wood-filled composite based on recycled polyethylene terephthalate. Production and properties *International Polymer Science and Technology* **42** pp 41-43